Confocal Microscopy and Multiphoton Excitation Microscopy

The Genesis of Live Cell Imaging

Confocal Microscopy and Multiphoton Excitation Microscopy

The Genesis of Live Cell Imaging

Barry R. Masters

SPIE PRESS

Bellingham, Washington USA

Library of Congress Cataloging-in-Publication Data

Masters, Barry R.
Confocal microscopy and multiphoton excitation microscopy : the genesis of live cell imaging / Barry R. Masters.
p. cm.
"Press monographs v. PM161"—Provided by publisher.
Includes bibliographical references and index.
ISBN 0-8194-6118-0 (alk. paper)
1. Confocal microscopy. 2. Multiphoton excitation microscopy. I. Title.

QH244.M37 2005
502'.82—dc22

2005026105

Published by

SPIE—The International Society for Optical Engineering
P.O. Box 10
Bellingham, Washington 98227-0010 USA
Phone: +1 360 676 3290
Fax: +1 360 647 1445
Email: spie@spie.org
Web: http://spie.org

Printed in the United States of America.

The International Society for Optical Engineering

Cover image: Confocal microscopy of a fixed, stained, vertical section of human skin. This is a biopsy specimen from the upper arm. The horizontal field width is 1400 μm.

**To our teachers who taught us,
so that we can teach others**

On looking back to this event, I am impressed by the great limitations of the human mind. How quick are we to learn, that is, to imitate what others have done or thought before. And how slow to understand, that is, to see the deeper connections. Slowest of all, however, are we in inventing new connections or even in applying old ideas in a new field.

Frits Zernike, Nobel Lecture, December 11, 1953

Contents

List of Abbreviations

AOTF	acousto-optical tunable filter
APD	avalanche photodiode
CCD	charge-coupled device
CRT	cathode-ray tube
CT	computed tomography
CSLM	confocal scanning laser microscope
DIC	differential interference contrast
DOF	depth of focus
DPH	diphenylhexatriene
FISH	fluorescence *in situ* hybridization
FLIM	fluorescence lifetime imaging
FRAP	fluorescence recovery after photobleaching
FRET	fluorescence resonance energy transfer
fs	femtosecond
GFP	green fluorescent protein
LED	light-emitting diode
LSCM	laser scanning confocal microscope
MIAM	multiple imaging axis microscopy
NA	numerical aperture
OCT	optical low-coherence tomography
PDT	photodynamic therapy
PMT	photomultiplier tube
ps	picosecond
PSF	point spread function
RMS	Royal Microscopical Society
SHG	second-harmonic generation
SNR	signal-to-noise ratio
SPAD	single-photon avalanche photodiode
STED	stimulated emission depletion

Preface

This text explains the fundamentals of confocal microscopy and multiphoton excitation microscopy. It presents the big picture of technological development in optical microscopy and provides insight into the origins, development, modification, and application of confocal and multiphoton excitation microscopes and their use in biology and medicine. This insight is presented in light of the key problems that each new invention attempted to solve, the various paths to the solution, the myriad interactions of various inventors and their associated technologies, and the practical limitations of each step of discovery and technological development. The human side of these technological developments is also revealed by describing the individual motivations that drove different scientists to their inventions, as well as the parallel developments that preceded each stage of technological development.

The repeated convergence of disparate techniques, instruments, theoretical studies, inventions, and reinventions from a wide variety of disciplines partially solved a series of problems in the field of microscopy and produced the current renaissance in modern optical microscopy. Innovative ideas and technical developments came from many individuals living and working in several countries around the world. Innovation evolves from a broad knowledge base, an awareness of advances in disparate fields of science, the courage to radically depart from mainstream thinking, and a clear understanding and statement of the problem to be solved. In many cases, innovations arise from technology transfer and not true invention.

Only recently have technical developments in many separate fields—for example, medical imaging and cell biology—spread across disciplines. There are many more examples of advances in different fields finding applications in optical microscopy. The field of digital image processing was first developed for air and space imaging applications. The field of adaptive optics, now being developed into optical microscopes and medical laser imaging devices, was first developed in the fields of astronomy and military laser weapons. Finally, the emerging medical imaging field of optical low-coherence reflectometry and tomography was first developed for the telecommunications industry as devices for checking fiber optics and integrated optical devices.

The biomedical applications of optical microscopy constitute an emerging field driven by spectacular advances in the field of *in vivo* microscopy. Advances in confocal microscopy are providing new and important technical solutions in the fields of endoscopy, minimally invasive surgery, dermatology, and ophthalmology. New technical advances in the fields of neurobiology and developmental biology build on the instruments described in this book. In many cases, the solutions to these problems required the optimization of one or more other solutions; typically, designs compromised one or more parameters (resolution, contrast, time for image acquisition) to serve a specific purpose. Optical microscopy began with the observation of living specimens, and recently there has been a revolution to return to the observation of *in vivo* specimens. The combination of spectroscopic techniques and optical microscopy has resulted in important advances in the field of "optical biopsy." Furthermore, these promising new diagnostic techniques are transitioning from the laboratory to the clinic.

There is an advantage to staying aware of the theoretical and technical advances of disparate fields of science. Such awareness may prove to be useful in the development of techniques that seem far from the problem at hand. Being aware of current and interesting problems in the biomedical area as well as advances in modern techniques of imaging, signal processing, nanotechnology, and integrated optics creates the conditions for success in interdisciplinary research. I hope the reader will find these themes useful for stimulating developments in new instrumentation as well as innovative, clinically useful applications in the exciting field of optical microscopy.

Since many excellent books, courses, and Internet resources are available that describe many aspects of modern microscopy, is there anything new to write on this subject? I delayed the completion of this book over a period of years while I attempted to answer this question myself. I believe the answer is a definite yes. In this book, I present several new ways of approaching these two topics. First, I present the recent developments as partial solutions to existing long-term problems. Second, I show that many developments are advances on previous instruments and techniques; there was an intellectual lineage in the development of modern microscopes. Third, I connect the developments of unique types of microscopes in disparate fields of science and medicine, and demonstrate their similarities. Fourth, I indicate the problems, limitations, artifacts, and experimental difficulties with modern microscopes. And fifth, I describe the techniques that use optical microscopes for studies on living tissue and organs and explain why the new types of microscopes are emerging as important clinical tools for medical diagnostics. *In vivo* microscopy and optical biopsy are active fields of research. This is evident from the exciting research in developmental biology, ophthalmology, dermatology, oncology, and brain imaging.

Another unique feature of this text is the discussion of the historical developments of optical microscopy and the technology's critical impact on the fields of biology and medicine. The reader may ask why this is necessary for an understanding of the modern instruments and their applications. There are several reasons. First, an appreciation of the chain of invention may serve to correct some incorrect attributions of priority and rediscovery of previous inventions. Second, an understanding of the historical development of both instruments and techniques has an important educational value in demonstrating serial and parallel approaches to problem solving in optical microscopy. Third, the study of the antecedents to various technical developments can put each invention and advance in its proper perspective and perhaps stimulate innovation. So many excellent books focus on applications that I decided it would be redundant to present extensive reviews of applications. Application notes are available online from the companies that manufacture confocal and multiphoton microscopes.

The projected audience for this text includes those who wish to gain insight into confocal microscopy and multiphoton excitation microscopy, and who intend to apply these techniques to biology and medicine. Therefore, it cannot be stated that this book was written for a single group of individuals. The projected audience includes undergraduate students who seek a global insight into the field of modern optical microscopy, graduate and postdoctoral students who will work with these instruments, and physicians who work with engineers and scientists to design and develop new, noninvasive, diagnostic instruments based on confocal or multiphoton excitation microscopy.

Optical microscopy is a nexus of theory, techniques, and devices from a wide variety of sources and disciplines, and the organization of this book reflects this fact. The text is divided into four parts. The largest part is devoted to confocal microscopy, with an introduction and a part devoted to multiphoton excitation microscopy. The emphasis is not on the main types of optical microscopes, but on how various technical developments served to solve the common problems of optical microscopy. However, throughout the book there are common themes, connections, and technical solutions to the problems of light microscopy that necessitated the deliberate repetition of some central concepts and ideas.

Each chapter of this text begins by introducing the materials to be covered and explaining their role in the book. A summary of key points at the end of each chapter reinforces those critical points. Hopefully the text contains everything that is essential and excludes those topics and details that are not critical for an understanding of the principles and their applications in microscopy. Further insight into the theory and practice of optical microscopy may be garnered by perusal of the printed and electronic resources that I have recommended in the appendix.

Part I covers the background, significance, and principles of the optical microscope. Chapter 1 presents a history of the microscope and the development of fluorescent microscopy, and describes the role of microscopy in the advancement of biology and medicine. Chapter 2 introduces the reader to the optical microscope by describing its chief components and limitations. Chapter 3 describes the contributions of Abbe, including the Abbe theory of image formation in an optical microscope. Chapter 4 discusses optical resolution in a microscope. When the major problems of optical resolution and optical aberrations were solved, the new primary problems concerned the development of techniques to provide contrast. These techniques resulted in the emergence of live cell imaging in optical microscopy. Chapter 5 explains both the nonoptical and optical techniques (phase contrast and differential interference contrast microscopy) that provide contrast.

Part II describes the partial solutions to the following problem: how to image thick, highly scattering specimens with an optical microscope. The invention of the confocal microscope, with its many technical variants, provided one partial solution. Confocal microscopy improves the resolution, contrast, and optical sectioning capability of the light microscope. The connecting theme in Part II is that a variety of techniques were invented and reinvented to solve the same problem: how to construct an optical microscope that has depth discrimination, and thus provide a microscope with the capability to "optically section" thick, scattering specimens. Chapter 6 formulates this problem and then describes several early antecedents to the development of confocal microscopy. Chapter 7 analyzes the myriad solutions to the problem of depth discrimination: the various types of confocal microscopes and their limitations.

Chapter 8 describes the development of scanning-slit confocal microscopes, which were developed in disparate fields: ophthalmology, neurobiology, and cell biology. Chapter 8 also plays a special pedagogical role in this book. While superficially it may seem that the theme is of interest only to ophthalmologists because the applications are predominantly imaging of the *in vivo* eye, there is a much deeper motivation to include these topics. This chapter demonstrates the linkages, connections, and technology transfers from numerous sources in the progression of technological development of the confocal microscope. For example, the inventions of Baer were motivated by the desire to develop a confocal microscope for cell biology, and the inventions of Svishchev were motivated by the desire to develop a confocal microscope to study neurobiology.

The primary message contained in Chapter 8 is that technical problems are solved by building on the previous and parallel work of others. The insights exposed in this chapter were derived not only from reading the published papers and patent literature, but also from personal conversations with Maurice, Svishchev, Petràn, Hadravsky, Baer, Koester, Kino, and Thaer. I also gained insight from working in the laboratory with Kino, Maurice, and Thaer. This chapter also provides an important lesson: teachers should teach not only those techniques that are popular; they must have a larger objective to teach how to solve problems by devising creative solutions. Many of the technical advances developed in Chapter 8 have found their applications in modern biomedical confocal instruments: scanning-slit confocal microscopes to investigate the cochlea, study *in vivo* human skin, and study the normal and pathological eye. Confocal microscopes based on slits are also being developed to image large embryos and study their development.

Chapter 9 describes the components of a confocal microscope. Even with the invention and development of the many types of confocal microscopes, problems remain. First, the ultraviolet excitation light used to excite many fluorescent dyes in molecular biology, ion indicator dyes, and endogenous molecules such as NAD(P)H and neurotransmitters, with absorption bands in the ultraviolet, is toxic to live cells, tissues, and organisms. Second, the depth of penetration of thick, scattering specimens is a few hundred microns and therefore precludes the imaging of thicker specimens. Third, the highly intense visible and short wavelength light causes photobleaching of the specimens during observation. The partial solution to these problems came with the invention and development of nonlinear microscopy.

Part III describes nonlinear optical microscopy with an emphasis on multiphoton excitation microscopy. Chapter 10 presents the development of nonlinear spectroscopy and microscopy—in particular, the seminal role played by the invention of the laser. Chapter 11 presents a detailed description of multiphoton excitation microscopy, from the Göppert-Mayer theory (Maria Göppert, 1929) to the Denk,

Strickler, and Webb 1990 *Science* publication. Chapter 12 summarizes the theory behind and describes the instrumentation of multiphoton excitation microscopy.

Part IV discusses the path to imaging live cells, tissues, and organs. Chapter 13 sets out the remaining problems and describes the limitations of nonlinear microscopy. Chapter 14 presents future directions for confocal and multiphoton excitation microscopy. Chapter 15 addresses the important topic of laser safety and includes a section on how to clean optics. An epilogue discusses humans as tool makers and tool users.

The book concludes with an appendix containing an annotated listing of carefully selected reference materials and resources. They present applications in great detail as well as experimental protocols. The appendix also contains a partial listing of the author's publications in ophthalmology and dermatology that illustrate the benefits of confocal and multiphoton microscopy in clinical medicine.

This book tells the story of the development of solutions to formidable problems in optical microscopy. It also tells the story of the limitations of optical microscopy: optical aberrations, optical artifacts, fundamental physical limitations of signal and noise, the quantum nature of light, stray light, background fluorescence, and light damage to the specimen. The information in this book will be an ongoing story—microscope development continues as an active field of progress toward the partial solution of the following problems: resolution, contrast, and optical microscopy of live cells, tissues, and organisms with minimal toxic and destructive effects. There is much work to be done, as we have only partial solutions to these problems. The state of the art is a moving target.

Finally, I gladly thank Margaret Thayer and Sharon Streams of SPIE for their help with the manuscript.

Confocally yours,

Barry R. Masters
November 2005

Part I

Optical Microscopy

Chapter 1

A Brief History of the Microscope and its Significance in the Advancement of Biology and Medicine

This chapter provides a historical foundation of the field of microscopy and outlines the significant discoveries in the fields of biology and medicine that are linked to the microscope. Microscopes, which are devices to image those objects that are invisible to the naked eye, were transformed from interesting instruments used by hobbyists to serious scientific instruments used to explore and understand the microscopic world. Because the technique of fluorescence microscopy is a major, if not the most widely used, application of both confocal microscopy and multiphoton excitation microscopy, I present a series of key developments of fluorescence microscopy. Microscopy began with the observation of live specimens and continues its growth with technical developments in the fields of intravital microscopy, endoscopy, and *in vivo* microscopy. In this chapter, I cite and discuss many of the advances in both biology and medicine that critically depended on the development of the optical microscope. These sections provide a framework for the book and support the premise that technical advances in microscopy have led to both the generation of new knowledge and understanding as well as advances in diagnostic and clinical medicine, which has ultimately resulted in an improvement of the human condition.

1.1 Timeline of Optical Microscope Development

The invention of the microscope (ca.1600) and its improvements over a period of 400 years has resulted in great advances in our understanding of the microscopic world as well as extremely important advances in biology and medicine. The optical microscope, a device that in many cases was used as an interesting toy, became a key instrument in basic science and clinical research: it gives the observer a view of inner space, that is, the world that cannot be observed with the naked eye because of insufficient resolution, such as atoms, molecules, viruses, cells, tissues, and microorganisms.

The reader may ask, why were the numerous early advances made in the design and manufacture of telescopes not rapidly transferred to the microscope? A partial answer is that telescopes were the domain of physicists and mathematicians, whereas the design, construction, and use of the early optical microscopes were left to laypersons, those whom today we call hobbyists. As we shall see, there were brilliant exceptions, and the application of mathematics and physics ultimately had great impact on the development of optical microscopes.

The history of the microscope is intimately connected with advances in optics. Advances in optics took place over hundreds of years, with contributions from scholars in many lands. One outstanding example is the work of Abu Ali al-Hasan ibn al-Hasan ibn al-Haytham, also known as Ibn al-Haytham or Alhazen. He was a Persian mathematician and astronomer who worked in Cairo, Egypt. Ibn al-Haytham wrote his treatise *Kitab al-Manazir* ("Optical Treasures") in the second quarter of the 11th century A.D. The first Latin translation, which reached Europe at about 1200 A.D. was called *Perspectiva* or *De aspectibus*; in 1572 a Latin version was printed in Basel with the title *Opticae Thesauris*. He described the laws of rectilinear propagation of light and of reflection and refraction. In the late 16th and 17th centuries, the *Opticae Thesauris* was known to Willebrord Snellius (who by experimentation rediscovered the law of refraction), René Descartes, Johannes Kepler, and Christiaan Huygens.

The laws of reflection and refraction were used to design optical instruments for many years, but after the 1690 publications of Huygens' *Traité de la Lumière,* the Huygens construction was used to trace geometrical wavefronts. The seminal work of Abbe applied the wave properties of light, specifically light diffraction, to image formation and optical resolution in the light microscope.

In the 17th century, advances in optics such as the law of refraction, geometrical optics, ray tracing, and Huygens' theory of light contributed to advances in microscopy. In the 19th century, the theory of diffraction was exploited by Abbe to explain optical resolution in a microscope. In the 20th century, the theories of interference and light polarization were developed into the interference microscope and the phase contrast microscope. These technical advances resulted in optical microscopes that provide contrast in living, unstained cells and tissues. The development of the electron microscope built on advances on the understanding of wave optics, especially in the design and construction of magnetic lenses to focus the electron beam.

Many books in many languages are devoted to the historical development of the optical and electron microscope. The World Wide Web also contains several very interesting websites devoted to the history of the microscope, which can be easily found by means of a search engine such as Google. The following is a brief timeline of some of the microscope developments and findings that resulted from advances in microscopy. What follows is neither comprehensive nor complete; it is only a brief survey of some of the many points of interest. If you are stimulated to further explore these fascinating topics, then please continue your learning with some of the excellent books, papers, and websites devoted to them.

In 1590 the Dutch spectacle makers Johannes Jansen and his son Zacharias produced a microscope based on two lenses held within a tube. In 1667 Robert Hooke published his book *Micrographia*, which included his many wonderful observations of the microscopic world. He correctly described the fruiting structures on living molds. The drawings of these microscopic observations contained in his book helped promote rising public interest in microscopy.

In 1675 Antony van Leeuwenhoek, a cloth merchant in Delft, constructed a single-lens microscope (see Fig. 1.1). He used a small double-convex lens with a maximum magnification of about 270×. The source of illumination was the sun, and the eye was the light detector. Leeuwenhoek observed and reported on bacteria, spermatozoa, red blood cells, simple plants, the structure of the cornea, the ocular lens, the optic nerve, the cornea, and striated muscle. Thus began live cell imaging with the microscope.

In 1830, Joseph Jackson Lister demonstrated how a combination of several lenses could minimize the problem of spherical aberrations. He used one lens with a small spherical aberration and then added a series of lenses to form a high magnification from the entire set. The additional lenses do not add to the spherical aberration of the first lens, but they increase the total magnification. This important advance allowed the objective to be constructed with increased apertures, which resulted in increased resolution.

Another difficult problem was chromatic aberrations. In 1813, the Italian botanist Giovanni Battista Amici solved this difficult problem by inventing a horizontal achromatic reflecting microscope based on mirrors. Later, in 1850, he used water-immersion microscope objectives that had improved resolution. In 1816, Fraunhofer invented a single achromatic lens that consisted of two different glasses in contact. Until the 1830s, with the development and wider availability of achromatic

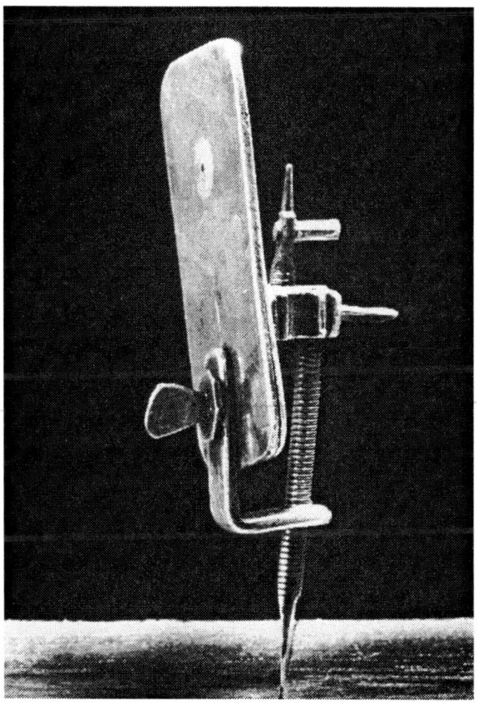

Figure 1.1 A typical Leeuwenhoek single-lens light microscope.

Figure 1.2 Ernst Abbe.

microscopes, the optical quality of microscopes did not surpass the quality of images obtained with the simple, single-lens microscope!

Following the inventions of Lister and the subsequent solution offered by achromatic microscope objectives, the next important problem was to increase the resolution of the optical microscope. In the 1870s Ernst Abbe (see Fig. 1.2) in Jena worked out the diffraction theory for image formation and derived a formula (Abbe formula) that related resolution to the wavelength of the illumination light and the numerical aperture (NA) of the lens. Abbe showed that in order to maximize the resolution of the microscope, it is necessary to collect as large a cone of light from the specimen as possible. Chapter 3 further describes the development of the Abbe resolution formula.

Although the principle of immersion microscope objectives was known for 200 years, Abbe began 10 years of work on the design of new immersion objectives in 1878. In the 1890s, he introduced several oil-immersion microscope objectives with a NA of 1.4, which were incorporated into the Carl Zeiss microscopes (see Fig. 1.3). These newly developed optical microscopes achieved their theoretical resolution of 0.2 μm with visible light.

Another productive scientist at the Carl Zeiss Corporation was August Köhler (see Fig. 1.4). In 1893, Köhler invented the subsequently named Köhler illumination system for microscopes. This important advance permitted uniform illumination of the specimen, as well as offering the highest obtainable resolution. Today, all commercial light microscopes are designed for Köhler illumination. Köhler illumination is described further and illustrated in Sec. 2.7.

Figure 1.3 Carl Zeiss.

Figure 1.4 August Köhler.

Critical illumination uses a light source such as a filament lamp followed by a field stop. The light passes through an aperture stop and then onto a condenser lens. The aperture stop sets the NA of the condenser lens. The light from the condenser lens is directly focused on the specimen.

With **Köhler illumination,** a different optical setup is used that provides uniform illumination of the specimen in the object plane. A lens and a field stop image the light source onto the back focal plane of the condenser, which provides uniform illumination in the object plane. Note that critical illumination is much brighter than Köhler illumination; however, it is very uneven, especially with low-power microscope objectives.

Another important technical advance derived from Abbe and incorporated by Zeiss into its microscopes was the Abbe microscope condenser, which is a common form of the bright-field condenser. It was constructed from two single-lens elements. The Abbe microscope condenser is designed to have a NA large enough to match that of any achromatic microscope objective.

Abbe invented, designed, and constructed new optical instruments and components for the light microscope; furthermore, he developed a theory of the light microscope and performed experiments to validate his theory. The **Abbe diffraction theory** of image formation, several methods for forming contrast, and a discussion of various definitions of resolution in the optical microscope will be discussed in subsequent chapters. Between 1888 and 1895, Abbe published a series of articles describing his complete theory of image formation in the optical microscope. With the use of apochromatic objectives and the homogeneous immersion technique, the optical microscope achieved an Abbe resolution of about 0.2 μm. In fact, the Abbe resolution limit depends on many physical parameters and will vary under different circumstances: wavelength, NA, coherence of light. These advances permitted the observation of many types of bacteria.

In 1903 Richard Zsigmondy (Nobel Prize recipient in 1926) and Wilhelm Siebenkoph, while working at Carl Zeiss in Jena, invented the ultramicroscope. Basically, **ultramicroscopy** is a form of dark-field microscopy using a very bright source of illumination that is perpendicular to the optical axis of the microscope. The **optical axis** is defined as a straight line joining the centers of curvature of lens surfaces. Ultramicroscopy can detect colloidal particles that are much smaller than the calculated classical limit of resolution in an optical microscope. While these particles can be detected by the ultramicroscope, they are not resolved!

In 1911 and afterwards, all the microscope objectives made by Zeiss were parfocal. **Parfocal objectives**, which comes from a suggestion from Köhler, means that the image remains in focus when the observer changes one microscope objective for another. This advance makes it easy to work with several different microscope objectives.

After the technical solution to the problems of resolution in the optical microscope arrived, the next set of major technical developments were solutions to the problems of contrast; i.e., how to produce contrast in thin, transparent specimens such as living, unstained cells and tissues, which have little inherent contrast. Microscopic observation of thin transparent living cells are phase objects and are difficult to observe under a standard light microscope; the main effect of the light in transmitting through the cells is to change the phase of the light by different amounts as it travels through various regions of the cells. Unfortunately the human eye cannot detect differences in phase; but it can detect differences in light intensity.

The solution to this problem was the work of Fritz Zernike. In 1932, Zernike (Nobel Prize recipient in 1953) invented a phase contrast microscope, which converts small differences in the phase of the light interacting with a specimen into corresponding differences in intensity that the human eye can detect. This important invention resulted in the widespread application of the phase contrast microscope to the field of cell biology; in particular, to the microscopic observation of living cells in tissue culture.

In 1953, the French physicist Georges Nomarski invented the differential interference contrast (DIC) microscope, which can image transparent cells and tissues. The DIC microscope converts gradients of phase of the light interacting with a specimen into intensity differences. This technique is very useful for the observation of unstained biological specimens and permits the observation of internal structures in transparent cells.

It should be clear that the invention of improved optical microscopes was a necessary but insufficient condition to lead to many advances in biology and medicine. In addition to the new microscopes, it was necessary to develop instruments and techniques for sample preparation. The techniques of tissue fixation, embedding, sectioning, and staining were critical to the success of the microscope. For example, the invention of the microtome in 1856 by Welcker, used to produce very thin sections, was crucial to many of the advances in anatomy and histology that are associated with the microscope. The microtome provided a technique to section soft tissue after fixation, thus opening the door for the observation of bacteria in tissues and the start of microbiology.

Other highlights in the area of sample preparation include the following. The optical microscope is inherently two dimensional; therefore, a three-dimensional specimen has to transform into two dimensions, i.e., a very thin specimen. This was achieved by producing thin smears or by mechanically cutting or sectioning a fixed, hardened specimen.

Paul Ehrlich wrote his dissertation in 1878 on the theory and practice of staining tissues with aniline dyes. In 1882, he developed his method for staining the tubercle bacillus. Ehrlich showed that dyes could be classified as basic, acid, or neutral. His work became the basis of future work in hematology and the staining of cells and tissues.

In 1884 the Danish physician Christian Gram invented what became known as the Gram stain. His method consisted of staining with gentian violet and potassium iodide, which results in differential staining or the ability to separate bacteria into two classes: gram-positive and gram-negative. The invention of what is known as the Golgi silver stain by Camillo Golgi in 1873 permitted the observation of single neurons within the complex nervous systems of animals. This technique was modified and exploited between 1877 and 1900 by Santiago Ramón y Cajal in his seminal, extensive studies on the histology of the nervous system. In more modern times, the development of the scanning electron microscope, together with freeze-etching and freeze-fracture techniques, resulted in the observation of the internal fine structure of cells and membranes. For a wonderful study of the role of staining in microscopy, the reader is referred to *History of Staining* (Clark and Kasten, 1983). With the exception of a brief review of fluorescence microscopy, these advances in tissue preparation are not discussed further in this book.

Once the problems of optical aberrations and optical resolution were sufficiently solved to permit the manufacture of optical microscopes with sufficient resolution to resolve bacteria, the next stage was to develop techniques and methods to provide improved contrast and specificity. One of the great advances in optical microscopy, used in both confocal and multiphoton excitation microscopy, is the invention of fluorescence microscopy.

1.2 Key Developments of Fluorescence Microscopy and its Limitations, Genesis, and Some Applications

This section integrates the genesis of the fluorescence technique with its physical basis and points out some important applications. Fluorescence microscopy is a means to achieve high specificity and contrast. For example, using fluorescent techniques it is possible to label single proteins, single-cell organelles, cytoskeleton structures, cell membranes, parts of chromosomes, and single neurons; to monitor intracellular ion concentrations, transmembrane potential differences in excitable tissues, the expression of specific genes, and detect single molecules.

It is valuable to briefly review the origins of fluorescence spectroscopy, since this is the foundation of fluorescence microscopy. I strongly recommend that the reader who wishes to exploit the many aspects of fluorescent microscopy (single photon or multiphoton) become familiar with Lakowicz's excellent book *Principles of Fluorescence Spectroscopy* and also the catalog provided by Molecular Probes, Inc.

In 1838, David Brewster observed the phenomenon that today we call fluorescence. The great utility and specificity of fluorescence techniques in microscopy is related to two fundamental properties observed in Cambridge by George G. Stokes in 1852. Stokes, a physicist and professor of mathematics, observed what he coined "fluorescence" from a solution of quinine. The source of excitation was sunlight, the excitation filter was the colored glass of the church window, the emission filter was a colored glass of wine, and his eye was the detector. Stokes observed that the fluorescence typically is observed at longer wavelengths than the excitation light; consequently, today we label this effect the **Stokes shift**. It is because of the Stokes shift that sets of fluorescent filters can be used to isolate the fluorescence light from the excitation light. Stokes performed many experiments with the sun as the source of excitation light and liquid excitation filters to isolate the ultraviolet light. He used a yellow barrier filter made from a solution of potassium dichromate to separate the fluorescence from the excitation light.

The second property of fluorescence that is extremely useful in microscopy is that the absorption and emission of light from a fluorescent molecule is related to its structure. The Stokes shift varies for different fluorescent molecules; therefore, different fluorescent molecules can be used in parallel with different fluorescence filter sets. Fluorescent probes can be designed to cover the spectrum of available light sources. Modern confocal or multiphoton excitation microscopes can simultaneously image two or three different fluorescent channels, i.e., two or three different types of fluorescent molecules can be imaged simultaneously in the specimen.

Stokes and others observed that many natural substances, such as chlorophyll, show fluorescence. Autofluorescence was documented in 1911 by Hans Stübel, who investigated the natural fluorescence of teeth, bacteria, protozoa, proteins, and hemoglobin. Over the next several decades, the natural fluorescence of porphyrin breakdown products, lipofuscin, elastin fibers, and, more recently, the natural fluorescence of the cornea, ocular lens, and human skin were observed.

Fluorescent probes, stains, and intravital dyes also have a fascinating history. The development of these stains and fluorescent probes is integrated with the advances of microscope instrumentation for fluorescence microscopy. Haitinger in 1931 coined the word fluorochrome for a fluorescent stain that induces secondary fluorescence in tissues. One early example is the molecule fluorescein. It was first synthesized by Baeyer in 1871. It is of interest to note that in 1882 Paul Ehrlich used fluorescein to study the pathways of the aqueous humor in the animal eye. This may be the first reported used of an intravital dye in physiology.

Perhaps a precursor of the fluorescence microscope was the ultraviolet microscope developed at the Carl Zeiss factory by August Köhler at Jena, Germany. Shortly before that development, Köhler and Moritz von Rohr developed quartz monochromatic ultraviolet microscope objectives that transmit at 275 and 280 nm. From the first fluorescence microscopes by Köhler and Siebenkoph, to more advanced models by Carl Zeiss in Germany and Carl Reichert in Vienna, fluorescence microscopes gained performance and utility. Major technical advances included the development of objectives for the ultraviolet, new powerful light sources, and new types of excitation and emission filters. There were immediate successful applications in medicine. For example, Ehrlich used the fluorescence microscope and fluorescence dyes to observe bacteria in tissues.

In 1911, Hans Stübel used an ultraviolet fluorescence microscope to demonstrate cell damage caused by short-wavelength excitation light. He described the ultraviolet-light-induced death of paramecia. The problem of phototoxicity and photodamage is still a major limitation of *in vivo* microscopy for both the confocal and the multiphoton microscope.

In 1929, based on the work of Phillip Ellinger and August Hirt, Carl Zeiss produced a fluorescence microscope. Known as an intravital microscope, it used a water-immersion microscope objective, an ultraviolet light source, filters, and a vertical or epi-illumination system. The intravital microscope was used for studies of kidney function, liver function, and the detection of vitamins and bacteria in living tissues. Ellinger used the device to investigate the structure and alteration of the microvasculature.

Following the development of the new Zeiss microscope, Leitz in Germany produced what was called the Ultropak. This microscope was used for many studies of the intravital fluorescence of living organisms. Other manufacturers, such as Reichert in Germany and Bausch & Lomb, manufactured fluorescence microscopes.

What were the key technical advances that led to the widespread use of fluorescence microscopes in biology and medicine? In 1959, E. M. Brumberg published a paper, "Fluorescence microscopy of biological objects using light from above"(Brumberg, 1959). He described a special illuminator with interference dividing mirrors to separate the excitation light from the fluorescence emission. Brumberg's invention was further developed by J. S. Ploem to form the interchangeable dichroic mirrors that are used in fluorescence microscopes with incident light illumination.

In 1946, Larionov and Brumberg observed living mammalian cells with a reflected light microscope that used an ultraviolet light source. They observed that the appearance of living mammalian cells differs from that of injured or dead cells. This indicates the importance of live cell imaging with the light microscope.

Brumberg's reflected light fluorescence microscopy is an example of a **conventional epi-illumination microscope**. The entire field of view is simultaneously illuminated (full-field or wide-field illumination), and fluorescence or reflections from the complete depth of the specimen are imaged. Since the fluorescence comes from all regions and not just the focal plane, the resulting image is degraded with blur and a loss of contrast. This epi-illumination system has important advantages over the transmission light fluorescence microscope: the full NA of the microscope objective is utilized, and fluorescence microscopy can be combined with Nomarski differential interference microscopy. The dichromatic beamsplitter or dichroic mirror reflects the incident light at 90 deg. through the microscope objective to the specimen. The microscope objective functions as both the condenser and the image-forming lens.

Specificity is one requirement of the development of fluorescent probes. The word **fluorochrome** was coined in 1934 by Max Haitinger to describe fluorescent dyes used to induce fluorescence in tissues. The invention of immunofluorescent probes by Albert Coons in 1941 was a major development in the field of fluorescence microscopy. Coons invented a method that could localize specific classes of proteins in cells by chemically attaching fluorescein to an antibody. The very high specificity of the antibody-antigen interaction is the molecular basis. The invention and development of immunofluorescence was a great advance for clinical medicine. Today a number of fluorescent probes exist, such as various types of green fluorescent proteins, that can be expressed by cells and used as markers of gene expression in the study of complex developmental processes.

Both single-photon fluorescence confocal microscopy and multiphoton excitation microscopy depend on, and take advantage of, important previous developments in fluorescence microscopy. Therefore, it is important to present an overview of the historical development of fluorescence microscopy and fluorescent probes and staining techniques in order to place this technique in its proper context. See Kasten (1989) for a more detailed account.

Microscope components such as dark-field illumination, dichroic mirrors, epi-fluorescence illumination systems, and intravital microscopy were all in use prior to the invention of confocal and multiphoton excitation microscopes. In addition, autofluorescence, fluorescence probes, fluorescence-linked antibody probes, and light damage of specimens during microscopic observations were well known. Many types of fluorophores are used in biological imaging. They include molecules that show autofluorescence (intrinsic or endogenous fluorescence) such as NAD(P)H and flavins. Another class of fluorescent molecules, called fluorochromes, is introduced into the specimen and results in extrinsic or exogenous fluorescence. Examples of the latter include molecular fluorescent probes, fluorescent antibodies, and green fluorescent proteins.

The second requirement of fluorescent probes to be used for live cell and organism studies with fluorescence microscopy is that the incorporation of the probe or its genetic expression in cells does not alter the normal structure and function of the cells. Finally, the fluorescent probe must not kill or damage the cells in the presence or absence of excitation light.

The field of **quantum dots** fluorescent probes is a very active area of research and development for application in both confocal microscopy and multiphoton excitation microscopy. Quantum dots are semiconductor nanoparticles composed of thousands of atoms with several unique properties that are exploited in their role as fluorescent probes. Quantum dots can be thought of as nanoparticles in which an electron is confined in a three-dimensional well. The use of quantum dots as fluorescent probes provides several advantages over organic fluorophores. First, quantum dots can be produced in a wide spectrum of emission wavelengths; the emission spectrum is a function of the size of the nanoparticles. By selecting nanocrystals of a specific size, it is possible to tune the emission wavelength. In addition, by selecting the materials (e.g., CdSe, CdS, InAs) as well as the particle size, a very wide spectrum of emission wavelengths can be obtained, which is extremely useful for bioimaging. Second, quantum dot fluorophores have emission bands that are narrower than those for organic fluorophores. Third, the fluorescence lifetime of quantum dots is of the order of hundreds of nanoseconds. This property is useful when time-gated detection is used to separate the emission from the quantum dots from the much shorter lifetimes of cell autofluorescence. Fourth, almost no photobleaching of the quantum dots occurs. However, the quantum efficiency of the quantum dots is low, which results in a low fluorescence intensity. Major developments include water-soluble quantum dot fluorophores, quantum dots linked to specific biomolecules, and the development of biocompatible quantum dot fluorophores for cells and tissues.

As with many microscopic techniques, at least two major limitations are associated with fluorescence microscopy. The first is **photodamage,** which is associated with the fluorescent probe and living cells and tissues. It was noted many years ago that living cells and organisms are more sensitive to ultraviolet light illumination in the microscope following the application of fluorescent probes. In recent times, the photophysics of this process has been exploited in the therapeutic technique of photodynamic therapy (PDT) for cancer.

The second limitation is **photobleaching**, which is associated with the destruction of the fluorescence molecules. Experimentally this is observed as the loss of fluorescence of a stained specimen following continuous illumination with ultraviolet light and also with visible light, which causes the fluorophore to fluorescence. The basis for this phenomenon is the photochemical transformation of the fluorescent molecule into another molecule that is not fluorescent. It has been found that oxygen plays an important role in this process; therefore, reducing the concentration of oxygen (not advisable for living cells and organisms) can mitigate, but not eliminate, photobleaching.

Wide-field fluorescence microscopy is a highly useful technique that has extremely high specificity. For very thin mechanically cut sections, the image of the specimen is sharp and shows high contrast. Its limitations become evident for thick, highly scattering specimens, such as *in vivo* human skin, intravital microscopy of tissues and organs, *in vivo* brain imaging, and whole, living embryos. For these thick specimens, the image is blurred and the contrast degraded because of the fluorescent and scattered light from above and below the focal plane that contributes to the image. A wide-field fluorescence microscope has no depth resolution. The signal remains a constant value as the degree of defocus is increased.

Parts II and III of this book explain how the development of confocal microscopy and multiphoton excitation microscopy have solved this limitation.

1.3 Key Advances in Biology and Medicine Made Possible with the Microscope

In 1939 Kausch and Ruska in Germany made the first photomicrographs of the tobacco mosaic virus. For the first time, it was possible to observe a virus. After 1945, the invention of the electron microscope provided the researcher with a resolution that could not be obtained with optical microscopes. This development led to the understanding of the fine structure of viruses, the cell and its organelles, the nucleus, cell membranes, and neuronal synapses. It is important to state that these observations were made on nonliving cells and tissues and therefore could not capture structural changes, e.g., cell division.

We now briefly review the role of the optical microscope in biology and medicine. We will select some of the highlights in the history of microscopy to illustrate the connection between the discoveries and the optical microscope.

One of the last advancements in our knowledge of anatomy made during the Renaissance was the 1628 discovery of blood circulation by William Harvey. He used a magnifying glass, which he called a multiplying glass, to study the pulsations of blood flow in small animals and in his studies of the structure of dissected hearts, lungs, and blood vessels. Since he used only a magnifying glass and not a light microscope, he could not resolve and therefore could not observe what we call capillaries. But in 1660 Marcello Malphigi discovered capillaries with his microscopic observations of frog lungs. He also made many original observations in studies of chick embryology and the structure of human organs, such as the liver and kidney. Malphigi also used the microscope to discover taste buds and their associated nerves.

Robert Hooke in 1664 described the plant cells in wood, and details of the flea and the louse. His book, *Micrographia*, awakened the interest of the general public. During the same period, Jan Swammerdam observed erythrocytes and the two-cell division of a frog's egg. The mammalian ovarian follicle was discovered by Reinier de Graaf in 1672. From 1650 onwards, the light microscope was an important tool in the hands of anatomists.

The work of Leeuwenhoek stands out, not only because he built his own microscope, but also because he made many important observations: protozoa, striated muscle fibers, bacteria, spermatozoa, yeast cells, leukocytes, and the axon and myelin of nerve fibers.

With the development of the achromatization of the microscope and, hence, the correction of chromatic aberrations, another important set of medical advances occurred. In 1857, Pasteur discovered the lactic acid bacterium with an optical microscope. Another milestone that depended on the microscope was Pasteur's 1857 experiments that refuted the theory of spontaneous generation.

In the 1800s, the optical microscope was used in many studies of anatomy and histology. The concept of the cell is intimately linked with the optical microscope. The publications of Schleiden (1838) and Schwann (1839) on cell theory were other important milestones. About 100 years later in 1938, Rudolf Virchow published his book *Cellular Pathology*, which became the basis of the new science of pathology.

After 1878, when microscopes were developed with oil-immersion objectives, a series of important discoveries on the pathogenic nature of microorganisms occurred. The use of oil-immersion objectives increased the NA to about 1.4 and provided the maximum theoretical resolution with visible light. In the field of microbiology, Robert Koch used the microscope with the new Zeiss immersion objectives to discover the pathogenic protozoa and bacteria that caused tuberculosis and cholera, and the achromatic microscope permitted him to describe the life cycle of the anthrax bacillus.

In the 1880s, Eli Metchnikoff made important advances in understanding inflammation and the process of phagocytosis. The brilliant work of Ramón y Cajal on the structure of the nervous system took place over several decades and was made possible with the use of a Carl Zeiss microscope with a 1.4 NA oil-immersion microscope objective.

Spectroscopy was first applied to chemical analysis in flames and later combined with telescopes to analyze the light from stars. When spectroscopy was combined with the optical microscope, the result was enhanced chemical specificity and a long series of important advances in fluorescence microscopy.

The microscope was also used to study cells and tissues based on their absorption and emission spectra. The application of spectroscopy to medicine has a long, innovation-filled history. The light microscope was an integral part of instruments designed for both microabsorption studies and microfluorometric studies. The combination of the microscope and the spectrometer or fluorometer permitted the localization of the signal to specific regions of a cell. With the microscope, it became possible to measure the fluorescence from a specific organelle within a single cell—for example, to characterize the fluorescence of nucleic acids and nucleoproteins that were stained with acridine orange. Microfluorometric studies permitted the quantitative measurement of the autofluorescence from the mitochondria under a variety of physiological states.

In the late 1800s, Charles Alexander MacMunn investigated the spectra of heme proteins in different states of oxygenation. He summarized his spectroscopic findings in two important books: *The Spectroscope in Medicine* (1880) and *Spectrum Analysis Applied to Biology and Medicine* (1914). These early investigations were extended by David Keilin from 1925 to 1960, during which he used optical spectroscopy to study the respiratory chain and cytochromes common to plants, yeasts, and higher animals.

Otto Warburg in the early 1930s observed the fluorescence of NADH in solution. He used near-ultraviolet excitation light and observed the fluorescence at 460 nm. Warburg's work was seminal to later studies on the fluorometry of NADH in mitochondria and muscle.

In 1950 Torbjoern O. Caspersson of the Karolinska Institute, Sweden, published *Cell Growth and Cell Function, a Cytochemical Study*, which summarized his 20 years of research on microspectrophotometry of cell organelles, nucleotides, and proteins during the cell cycle, growth, and differentiation. Later, Rudolf Rigler, Jr., developed microscope-based instrumentation to study nucleic acids within cells using the technique of microfluorometry. There is a direct link among the works of Keilin on respiratory proteins, the prolific work of Caspersson on cellular microspectrophotometry, the microspectroscopy studies of Bo Thorell, the cellular fluorescence microscopy studies of Joseph Hirschberg, Elli Kohen, and Cahide Kohen, the work of Rudolf Rigler on cell microfluorometry, and the innovative studies of Britton Chance on the application of spectroscopic techniques to cellular respiration. Analytical cytology made great gains in Stockholm from 1945 to 1950.

The light microscope was initially used to explore the microscopic living world. Ancillary techniques such as fixing, mechanical sectioning, and staining were necessary components for its contributions in the life sciences and medicine. The development of the fluorescence microscope together with the continuing development of new, more specific stains and dyes resulted in tremendous gains in specificity and contrast. The invention of the phase contrast microscope and the differential contrast microscope permitted the observation of live cells and tissues; however, the long-term observations of thick, highly scattering tissues, embryos and organisms were still extremely difficult, if not impossible. *In vivo* microscopy began with Leeuwenhoek and continues today as a robust microscopic tool in the neurosciences, developmental biology, and as a clinical diagnostic tool in ophthalmology and dermatology.

In this chapter I have placed the development of the optical microscope in its important place in the history of biology and medicine. In Chapter 2, I present the principles and components of the optical microscope and discuss its limitations.

1.4 Summary

- The invention of the microscope (about 1600) and its improvements over a period of 400 years resulted in great advances in our understanding of the microscopic world and extremely important advances in biology and medicine.
- In 1816 Fraunhofer invented a single achromatic lens that consisted of two different glasses in contact.
- In 1830, Joseph Jackson Lister demonstrated how a combination of several lenses could minimize the problem of spherical aberrations.
- In 1878 Ernst Abbe in Jena worked out the diffraction theory for lens image formation and derived a formula (Abbe formula) for the maximum resolution in optical microscopes.
- With the use of apochromatic objectives and the technique of homogeneous immersion, the optical microscope achieved the Abbe resolution of about 0.2 μm, which permitted the observation of many types of bacteria.
- The brilliant work of Ramón y Cajal on the structure of the nervous system took place over several decades and was made possible with the use of a Carl Zeiss microscope with a 1.4 NA oil-immersion microscope objective.

- In 1929, based on the work of Ellinger and Hirt, Carl Zeiss produced a fluorescence microscope. This new fluorescence microscope used a water-immersion microscope objective, an ultraviolet light source, filters, and a vertical or epi-illumination system.
- The invention of immunofluorescent probes by Albert Coons in 1941 was a major development in the field of fluorescence microscopy.
- In 1959 E. M. Brumberg described a special illuminator with an interference dividing (dichroic) mirror to separate the excitation light from the fluorescence emission. Similar dichroic systems are used in all modern fluorescence microscopes.
- The invention of the electron microscope permitted the investigation of the fine structure of cells, synapses, and viruses. Its resolution, which exceeded that of light microscopy, is due to the shorter wavelength of the electrons in the electron microscope.
- The limitation of wide-field fluorescence microscopy becomes evident for the case in which the specimen is a thick, highly scattering object. For these specimens, the image is blurred and the contrast degraded because of fluorescent and scattered light from above and below the focal plane that contributes to the image. A wide-field fluorescence microscope has no depth resolution. The signal remains a constant value as the degree of defocus is increased.

Chapter 2

The Optical Microscope: Its Principles, Components, and Limitations

2.1 What is an Optical Microscope?

How does a slide projector differ from a microscope? A slide projector magnifies the image on the slide; hence, it projects a small image into a larger image on a screen. A slide projector does not increase the resolution of the object.

A microscope also provides a magnified image for the observer, although its most important function is to increase the resolution! With a microscope, we can observe microscopic specimens that would not be visible and resolve details that were unresolved to the naked eye. But unless there is sufficient contrast, no details can be observed. So, optical microscopy depends on both sufficient resolution and sufficient contrast.

2.2 Image Fidelity: Mapping the Object into the Image

As in all imaging systems, the optical microscope maps an object into an image. An ideal system would make this mapping with the highest fidelity between the object and the image. Even so, the finite aperture of the lens as well as many forms of optical aberrations place fundamental limits on the fidelity of this mapping. The aim of microscope design, manufacture, and practice is to minimize the aberrations, maximize the resolution, and approach the highest fidelity possible.

What are the requirements for **spatial and temporal resolution** in optical microscopy? Spatial resolution denotes the ability of the microscope to resolve or separate adjacent points on the object. Microscopic observations may only involve the detection or absence of a particle, or may require the full three-dimensional structure of a thick, highly scattering specimen such as the eye or skin. The microscope should be capable of resolving the highest spatial frequencies that are required to form an image that is appropriate to the questions posed by the observer.

In order to map the object into the image with high fidelity, it is necessary to map the intensities and the spatial frequencies of the object. **Spatial frequency** is the frequency in space for a recurring pattern, given in units of line pairs/mm. The **Nyquist theorem,** which is valid for both spatial and temporal frequencies, defines how to sample the object. The theorem states that the sampling must be performed at a minimum of two times the highest spatial frequency in the object to accurately reproduce the object in the image.

If the imaging system does not meet the Nyquist criterion, then there is aliasing in the image. **Aliasing** is the phenomenon that occurs when periodic structures in the object are not correctly mapped into the image; hence, the image has the incorrect periodic structure. Specifically, aliasing causes spatial frequencies higher than the Nyquist frequency to be displayed at lower frequencies. Aliasing is another form of artifact in the image.

Note that there is a trade-off between field of view and resolution. If we require a large field of view in the image, then the image will have a lower resolution compared with a high-power microscope objective and a high NA. Recently, some microscope manufacturers (Leica and Nikon) have produced new, non-Royal Microscopical Society (RMS) threaded microscope objectives that combine high NA (higher resolution) with a lower magnification. The area of the aperture in the back focal plane and the threaded diameter of the objective are increased over the standard RMS threaded diameter in order to manufacture these new microscope objectives.

In addition to the spatial resolution, the transverse resolution (in the plane of the specimen) and the axial resolution (along the optical axis of the microscope), there is also temporal resolution. If the specimen is fixed, nonliving, and stationary, then time is not a consideration. But if we are using the optical microscope to observe time-dependent events, e.g., changes in ion concentration, calcium waves in excitable tissue, alterations of intensity in live brain optical imaging, or cell and tissue changes in cell division, fertilization or embryo development, then temporal resolution is important. In general, we are required to acquire separate images (time sequence of images) that do not distort the temporal events observed. To do this correctly, the microscope should acquire images at a rate at least twice that of the most rapid process. This image acquisition speed will ensure that the time events are not distorted.

Finally, what contributes to a loss of image fidelity? First, if the resolution of the optical microscope is too low to image the fine details of the specimen, i.e., those parts with a high spatial frequency, then there will be a loss of fine details. Second, if the kinetics of the process under observation is too rapid compared to the image acquisition time, then the observed kinetics of the events will be distorted. Third, optical aberrations in the microscope can degrade the resolution of the image. Fourth, in order to achieve the maximum diffraction-limited performance of a microscope, it is necessary to use the microscope objective at its full NA. If the illumination source does not completely fill the back focal plane of the objective, the resolution will be compromised. Fifth, resolution is one requirement for image formation; appropriate contrast levels are also required.

The optical surfaces of all elements of the microscope, especially the objective, must be free from dirt, oil, dust, fibers, and mechanical scratches. Dust and mechanical scratches degrade the image quality, reduce resolution, and contribute to increased stray light with concomitant decrease in image contrast. Stray light in an optical microscope must be minimized since it also degrades contrast. In subsequent chapters, we will discuss other factors that reduce image fidelity, including photophysical bleaching of fluorescence of the specimen, illumination-induced cell and tissue death and damage, signal-to-noise ratios (SNR) and their effects on image detection, and the role of statistics and the quantum nature of light in light detectors.

The next section introduces the various types of optical aberrations. Although modern microscope objectives are available with high degrees of correction for various types of aberrations, the history of our understanding, measurement, and correction of optical aberrations represents a major achievement in the advancement of optical microscopy.

2.3 Optical Aberrations

Optical aberrations represent the failure of an optical system to produce a perfect image. They are the deviation caused by the properties of the lens materials or the geometric forms of the refracting and reflecting surfaces.

Modern microscope objectives are manufactured to minimize five categories of optical aberrations: spherical aberrations, coma, astigmatism, field curvature, and distortion. This set of monochromatic optical aberrations is called Seidel Aberrations in honor of Ludwig von Seidel, who classified them. The aberrations must be corrected in the listed order; i.e., to correct for astigmatism it is first necessary to eliminate spherical aberrations and coma.

The presence of **spherical aberration** results in the lack of a sharp focus point; instead there is a zone of confusion or caustic. This aberration is caused by a lens with spherical surfaces for which the peripheral regions refract light more than the central regions. The optimal correction for spherical aberration of a microscope objective requires a defined object and image distance. This explains why the results of high-NA, oil-immersion objectives used with a coverslip to image thick specimens are severely limited by the generation of spherical aberrations at increasing distances below the coverslip. Other sources of spherical aberration are mismatch of tube length and objective, nonstandard thickness of coverslips, and poor-quality immersion oil.

Coma is a lens aberration that occurs when light is focused at points off the optical axis. The **optical axis** is perpendicular to the plane of the lens and passes through the center of a circular lens. The name, derived from the Latin term for comet, is due to the fact that the aberrated image of a point looks like a comet.

Astigmatism must be corrected after spherical aberration and coma are corrected. The Seidel aberration of astigmatism is not equivalent to the term astigmatism as applied to human vision. For the human eye, the nonspherical shape of the lens results in different foci for different meridional planes. In contrast, Seidel astigmatism can occur with perfectly spherical lens surfaces.

It is first necessary to define two planes in the optical system. The **meridional or tangential plane** contains both optical axis and the object point. The **sagittal plane** is perpendicular to the tangential plane and contains the object point. What is observed is that points will be blurred only in a circular direction in the tangential focal plane. In the sagittal focal plane, only the radial direction has blurring.

Field curvature is another aberration that persists after spherical aberration, coma and astigmatism are corrected. In the presence of a lens with field curvature, object points that are in a plane will be imaged onto a paraboloidal surface. Field curvature makes a flat field appear curved and various regions of the image to be blurred. In the presence of field curvature, when imaging with a high-aperture microscope objective, one observes that either the center or the peripheral of the field of view is sharply focused.

Distortion is a displacement of the entire image rather than a blurring of the individual points that form the image. Distortion occurs when the lens magnification varies from the center to the periphery. Distortion can occur as either pincushion or barrel distortion.

In addition to the previous Seidel aberrations, corrections must be made for axial and lateral chromatic aberration, which cause the focus position to depend on the wavelength of the illumination light. Spherical and chromatic aberrations affect the entire field; in contrast, the other types of aberrations are only important for off-axis image points.

Axial chromatic aberration occurs when different light wavelengths are not focused at a single point on the optical axis. Each color of light will focus at a different point on the optical axis. The image is surrounded by fringes of different colors that change with varying focus. A concave lens of a glass of one refractive index can be joined to a second convex lens of a different refractive index to form an achromatic lens in which several wavelengths focus at the same point on the optical axis. Note the definition of refractive index: the ratio of the speed of light (phase velocity) in a vacuum to that in a given medium.

Lateral chromatic aberration occurs when different wavelengths are magnified at different ratios. This effect is greatest at the outside of the visual field of the object where the light rays are more oblique. Each object is surrounded by a colored fringe. This effect can be compensated by eyepiece design and the microscope objective (in older microscopes) or in the objective alone (in modern microscopes).

2.4 The Compound Microscope

The compound optical microscope uses two lenses (microscope objective and eyepiece lens) to project a magnified image of the specimen onto the image detector (solid state detector or the eye of the observer). Figures 2.1, 2.2, and 2.3 show the layout of the compound microscope, its conjugate planes, the illuminating light path, and the image-forming light path. The first lens is the microscope objective and the second lens is the ocular or eyepiece. Image formation interpreted in terms of the Abbe diffraction theory, to be discussed in the next chapter, is critically dependent on two lenses: the microscope objective and the condenser lens. The function of the microscope objective is to collect the light diffracted by the specimen and to form a magnified real image at the intermediate image plane near the ocular. The function of the condenser lens is to evenly illuminate the specimen.

It is necessary to define real and virtual images. A **real image** can be observed on a screen or captured on photographic film or on a solid state detector. In contrast

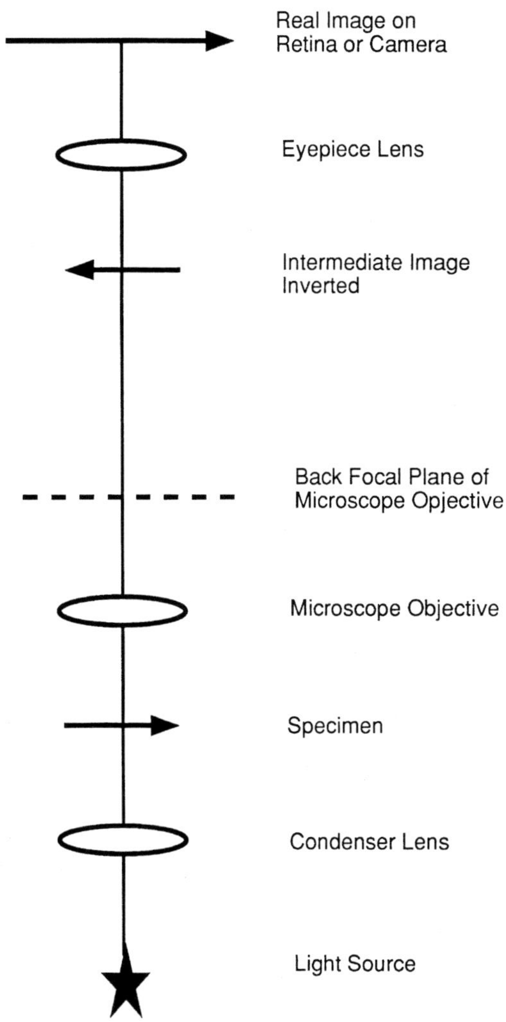

Figure 2.1 Schematic diagram of a compound microscope.

to the real image, a **virtual image** can only be observed or detected with the use of another lens or lens system.

2.5 Chief Components of an Optical Microscope

As shown in Fig. 2.1, the components of the microscope include the light source, illumination system, condenser, various diaphragms, the stage, immersion fluid, cover glass (or coverslip), microscope objective, tube, tube lens, eyepiece, and various filters, polarizers, and other optical elements. The detector (remember that intensity is detected) is either the naked eye or a film or electronic imaging system.

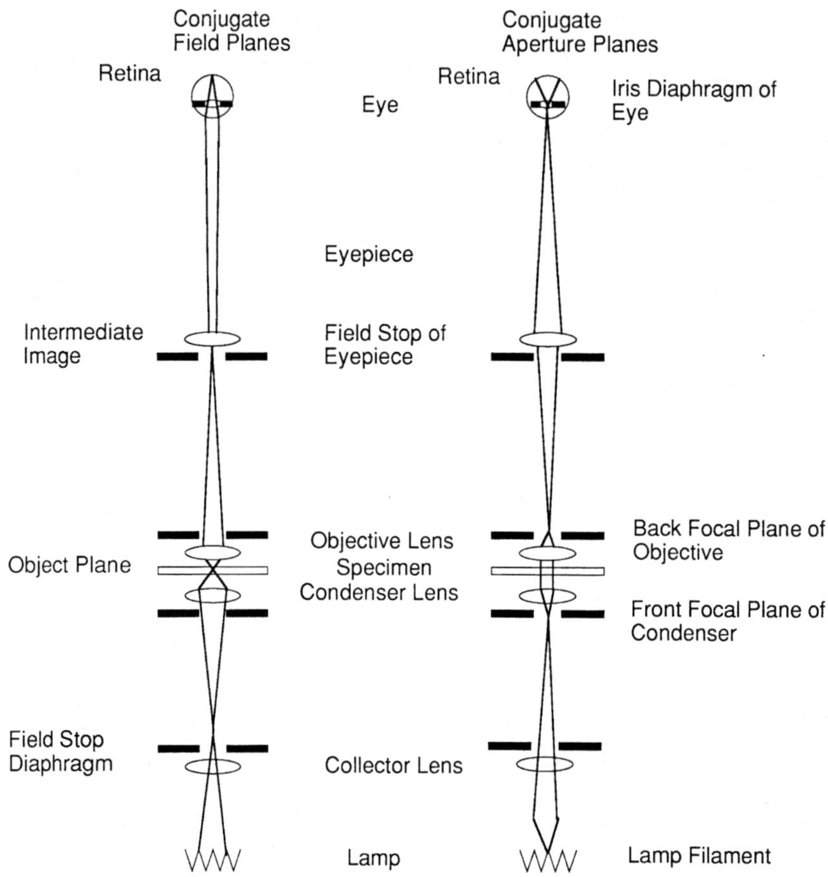

Figure 2.2 Schematic diagram showing the four conjugate field planes and the four conjugate aperture planes, showing bright-field Köhler illumination.

The light source illuminates the object and is the ultimate source of light that forms the detected image. The components located below the specimen include those of the condenser system. In addition to the condenser lens, there is a field diaphragm and an aperture iris. These components provide uniform illumination.

Above the specimen in Fig. 2.1 are two important magnifying lenses. The microscope objective is the most important optical element to form a high-resolution image; it collects the light from the various points in the specimen and redirects the light to the corresponding points in the image. A real image of the object is projected into the upper part of the microscope (microscope body tube). The second optical element, the eyepiece lens, forms a real image on the retina or the camera.

We begin by defining some useful terms. These terms and many others in English, German, and French are defined in the *RMS Dictionary of Light Microscopy* (1989).

An **aperture** is the area of a lens that is available for the passage of light. A **pupil** is defined as the apparent minimum common cross section of all light-ray bun-

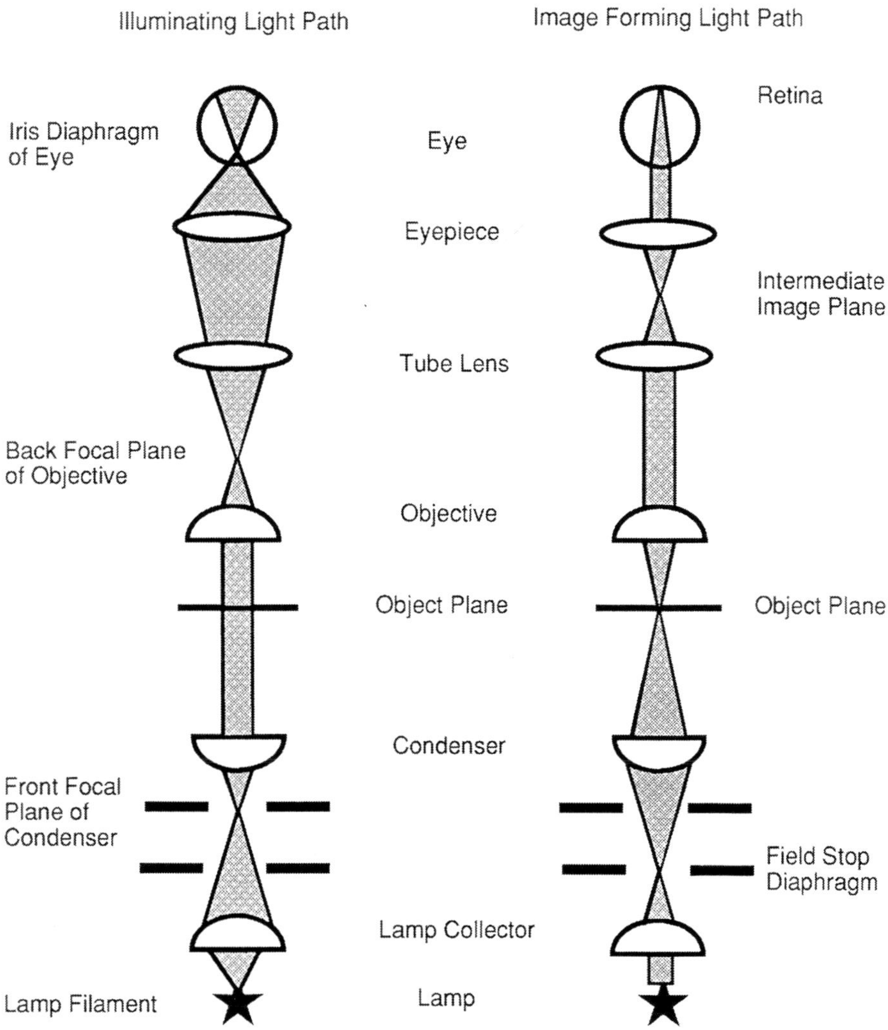

Figure 2.3 Schematic diagram showing the illuminating and image- forming light paths for an infinity-corrected light microscope with bright-field Köhler illumination.

dles both on the object side (the **entrance pupil**) and the image side (the **exit pupil**). A **diaphragm** provides a mechanical limitation of an opening normal to the optical axis that restricts the cross-sectional area of the light path at a defined place in the optical system. An **aperture diaphragm** is a diaphragm in the plane of any aperture of an optical system that limits its pupil (see Fig. 2.4). A **field diaphragm** is one in the object plane or any plane conjugate to it (see Fig. 2.4). The word **conjugate** means linked together by the rules of geometrical optics. It is also the name of the iris diaphragm in front of the collecting lens of the light source. With Köhler illumination, the condenser focuses the image of the field diaphragm onto the image plane (see Fig. 2.2).

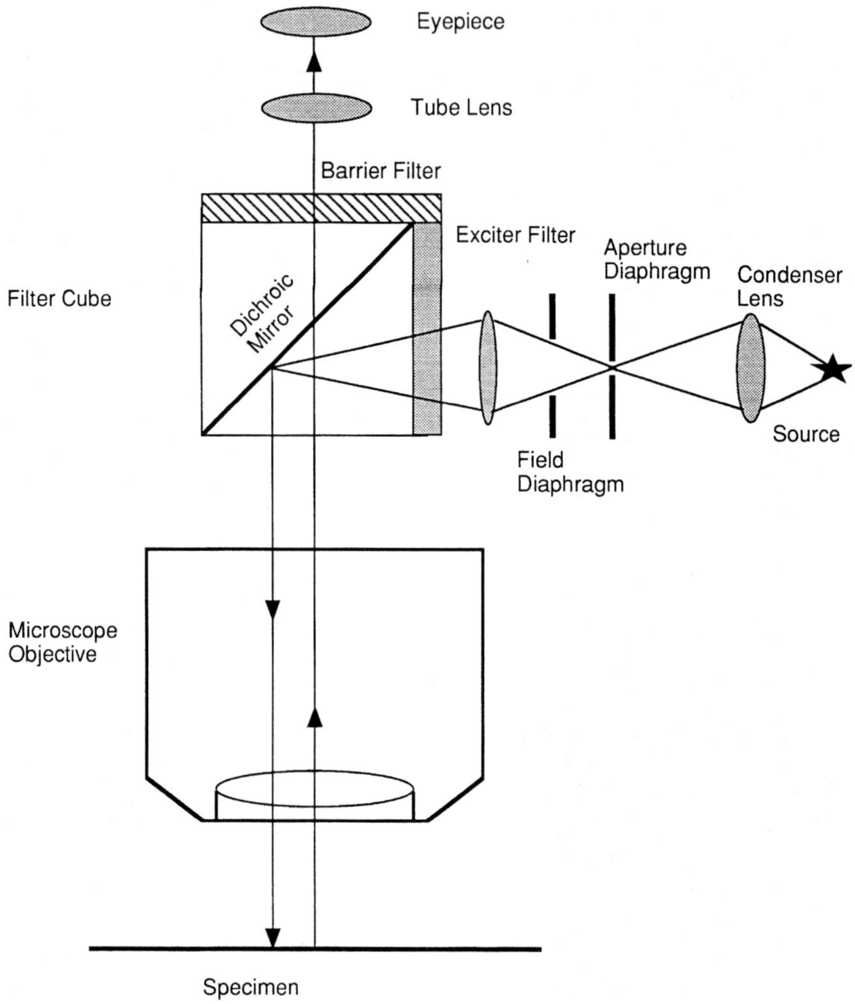

Figure 2.4 Schematic diagram of an epi-illumination fluorescence microscope.

A **condenser iris diaphragm** is located at the front focal plane of the condenser lens of a microscope. With Köhler illumination, the iris is located in a plane conjugate to the back focal plane of the objective lens. This iris continuously alters the numerical aperture of the condenser.

Dioptric describes optical elements, indicating that they operate by refraction, i.e., using lenses. **Catoptric** describes an optical system that operates by reflection. **Catadioptric** refers to an optical system with both reflecting and refracting surfaces that are used to form the image.

Two common terms used with lenses are back and front focal planes. The **back focal plane** of a lens is the focal plane that lies behind the lens when viewed in the direction of the passage of light. The **front focal plane** lies in front of the lens when viewed in the direction of the passage of light.

The mechanical parts of a classical optical microscope include the tube that contains both the microscope objective and the eyepiece, the stand that supports the complete instrument, and the specimen stage. The **tube** is that part of the microscope that connects the objective and the eyepiece.

The **mechanical tube length** is the distance from the top of the microscope objective housing to the top of the tube into which the eyepiece is set. For microscope objectives with short focal lengths, the mechanical tube length is approximately equal to the optical tube length that is defined as the distance from the back (rear) focal plane of the microscope objective to the intermediate image plane. Modern microscope manufacturers use different tube lengths: Nikon and Leica (200 mm), Olympus (180 mm), and Zeiss (160 mm).

Once the light microscope is properly aligned and adjusted, the spatial positions of all the components are fixed. Note that temperature variations and mechanical vibrations from the microscope system itself, or motors, controllers, shutters, lamp and laser cooling fans, or cooling water flow are greatly magnified and adversely affect operation. In addition, vibrations from sounds, elevators, motors, and street traffic can be transmitted to the microscope. Therefore, microscopes are usually placed on anti-vibration optical tables, and great efforts are made to isolate the microscope from all sources of vibration. The microscope stand is designed to provide a stable mechanical system to hold the various components.

The focal plane of the microscope objective is displaced through the volume of the specimen in two ways. First, the mechanical stage that contains the specimen can be manually or mechanically moved by a precise stepper motor driven along the optical axis of the microscope. Second, for very small but highly precise movements along the optical axis, a piezoelectric micropositioning device that is attached to the microscope tube on one side and the microscope objective on the other can be used to displace the objective relative to the stationary microscope stage. In each case, the distance between the specimen and the front lens of the microscope objective is changed. It is important to note that mechanical motion of any kind can and will induce transient mechanical vibrations in the microscope. Therefore, in an automated series of z-measurements or optical sections at different depths within the specimen, for example with the confocal microscope, it is important to allow mechanical vibration to be damped before each sequence of image acquisition.

The **eyepiece** or **ocular** is a lens system that is responsible for the angular magnifications of the final virtual image that it forms at infinity from the primary image. This is converted into a real image by the observer's eye or other converging lens system (see Fig. 2.2).

Another important optical element is the **tube lens** (see Fig. 2.3), defined as an intermediate lens designed to operate as an essential component of infinity-corrected objectives and located in either the body tube or the viewing tube of the microscope. The tube lens is associated with the objective lens system and will influence the effective magnifying power and possibly the state of correction of the system.

Finally, it is necessary to define two confusing terms: depth of field and depth of focus. **Depth of field** (depth of sharpness in object space) is the axial depth of the space on both sides of the object plane within which the object can be moved without detectable loss of sharpness in the image, and within which features of the object appear acceptably sharp in the image while the position of the image plane is maintained. **Depth of focus** (depth of sharpness in image space) is defined as the axial depth of the space on both sides of the image plane within which the image appears acceptably sharp while the positions of the object plane and of the objective are maintained.

2.6 Microscope Objectives

The microscope objective is a critical component of the optical microscope. Typically, a modern optical microscope contains several different objectives that are contained in a rotatable turret connected to the microscope tube. The sets of microscope objectives are **parfocal**; that is, they are mounted so that with the specimen in a fixed position each of the objectives is at the same level of focus within the specimen. This feature makes it easy to switch microscope objectives. A modern microscope objective is designed to minimize optical aberrations, stray light, and fluorescence from its components. Table 2.1 shows the various types of microscope objectives and their corrections.

Microscope objectives are optimized for specific applications and classified into broad groups with differences in the degree of correction from aberrations and also in cost. **Achromats** are objectives that are corrected at 540 nm for spherical aberrations. They are also corrected for chromatic aberration at both red and blue wavelengths (656 nm and 486 nm). They have excellent performance when used with monochromatic light. When used for low-magnification work, i.e., below 40x, they are a good selection based on performance and price.

Table 2.1 Microscope objectives and their corrections.

Type of Microscope Objective	Corrections for Aberrations		Field Curvature Correction Needed?
	Spherical Correction	Chromatic Correction	
achromat	1 wavelength	2 wavelengths	no
plan achromat	1 wavelength	2 wavelengths	yes
fluorite	3 wavelengths	3 wavelengths	no
plan fluorite	3 wavelengths	3 wavelengths	yes
apochromat	4 wavelengths	4 wavelengths	no
plan apochromat	4 wavelengths	4 wavelengths	yes

The word **plan** placed in front of the type of microscope objective indicates that the objective is corrected for field curvature. **Fluorite** or **semiapochromate** objectives are corrected for both spherical aberration and chromatic aberration at three wavelengths. These objectives can be used to the near-ultraviolet and have good contrast and high transparency. With these characteristics, they are useful for differential contrast, polarization, and immunofluorescence microscopy.

Apochromats, originally designed by Abbe in 1886, are the most expensive class of microscope objectives, but are highly corrected at four wavelengths for both spherical and chromatic aberrations. Useful for color microscopy with white light, they are available with large (1.4) NAs and are also transparent in the near-ultraviolet. Therefore, they are also very useful for low-light fluorescence microscopy and for fluorescence microscopy using dyes that have absorption bands in the ultraviolet region.

An **aplanatic lens** is corrected for both spherical aberrations and coma. **Achromatic** microscope objectives or achromats are corrected for spherical aberrations at one color (green) and for chromatic aberrations at two colors. **Apochromats** are corrected for spherical aberrations at two and for chromatic aberrations at three wavelengths.

A very useful technical advance is the advent of infinity optical systems. In a microscope based on a **finite optical system,** the light from a specimen passes through the objective and converges toward the primary image plane (see Fig. 2.5). The focus in the primary image plane is also the eyepiece focus point. That is the basic light path in a standard light microscope.

A microscope with an **infinity optical system** is very different. The microscope objective focuses light from a point source (i.e., a small fluorescent bead) to form a parallel beam of light. The infinity-corrected microscope does not form an image; instead the parallel beam of light is focused by a tube lens to form a real image in the primary intermediate image plane, which is conjugate with the object plane and the retina. In the space between the microscope objective and the tube lens (the infinity space), the light from the specimen is a set of parallel rays.

Many modern microscopes use **infinity-corrected microscope objectives**. An example, manufactured by Nikon for fluorescence microscopy, is the Plan Apo 60× oil-immersion, NA 1.40 with a free working distance of 0.21 mm. The **free working distance** is defined as the distance or depth of free space between the top or front lens of the objective and the surface of the specimen or the cover glass.

This microscope objective is designed to work optimally with a microscope that has a tube length of 200 mm, a thread diameter of 25 mm, and a parfocal distance of 60 mm. **Parfocal distance** of a microscope objective is the distance in air between the object plane (the uncovered surface of the object) and the locating flange of the microscope objective.

One advantage of infinity-corrected objectives is that it is simple to insert optical elements into the microscope tube (i.e., infinity space); for example, waveplates, filters, or compensators. The only optical requirement is that they have plane-parallel surfaces. In that case, their location is not critical and there is mini-

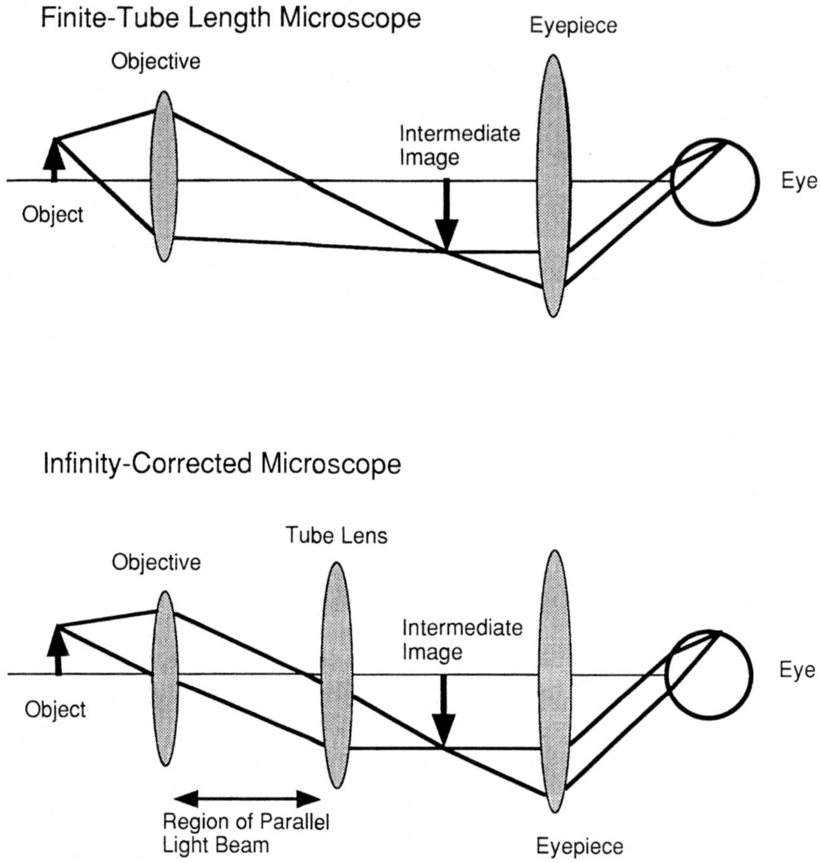

Figure 2.5 Schematic diagram of the ray paths for (a) a finite-tube-length microscope, and (b) an infinity-corrected microscope.

mal image shift, focus shift, or aberration as a result of their placement or removal. Another advantage is that a computer-controlled micropositioning device can displace the microscope objective with respect to a fixed stage and specimen without magnification error or aberrations. Some modern confocal microscopes use this feature to perform three-dimensional optical sectioning of a specimen. These two advantages have influenced manufacturers of modern optical microscopes to develop infinity-corrected microscopes that use infinity-corrected microscope objectives.

Usually a microscope objective is optimized for a specific use. The following characteristics are given by microscope manufacturers: the general class of objectives, e.g., plan fluor; the power, e.g., 100x; the NA, e.g., 1.3; the selection of immersion fluids, e.g., oil, water or air; the free working distance, e.g., 0.20 mm; the mechanical tube length; and whether the objective is infinity corrected as well as the thickness of the cover glass to be used. The higher the power of the microscope

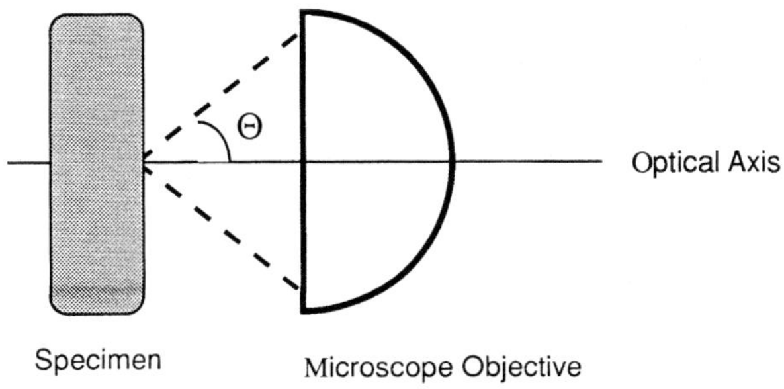

Figure 2.6 Schematic diagram showing the definition of Θ, which is the half-angle of the aperture cone of a microscope objective.

objective, the smaller the field of view of the specimen and the smaller the free working distance.

The NA of a microscope objective or a microscope condenser is defined as NA = $n \sin \Theta$, where n is the refractive index of the medium measured at 587 nm, and Θ is half the angular aperture, that is, the half-angle of the incident light rays to the top or front lens of the microscope objective (see Fig. 2.6). The **angular aperture** is defined as one-half the maximum plane angle subtended by a lens at the center of an object or image field by two opposite marginal rays when the lens is used in its correct working position. The symbol Θ is the half angle of the cone of light converging to an illuminated point or diverging from a point. This quantity can also be defined as the semiangle of the cone of rays from the axial object point that is received by the objective.

The **refractive index** is the ratio of the speed of light (its phase velocity) in a vacuum to that in a given medium. For an air or dry microscope objective, the medium is air with a refractive index of 1.000; for an oil-immersion microscope objective, the medium is oil with a refractive index of 1.515; and for a water-immersion objective, the medium is water with a refractive index of 1.333.

In general, a lens has two numerical apertures: one on the object side and one on the image side. In discussions of microscope objectives, the NA refers to the object side of the lens.

Three properties are a function of the numerical aperture. First, the higher the NA, the greater the resolving power, or the ability to resolve two points. Second, the higher the NA, the brighter the image in the microscope. Image brightness in transmission light microscopy is proportional to the square of the ratio of the NA to the total magnification. For the epi-illumination fluorescence mode, the image brightness is proportional to the fourth power of the NA divided by the square of the magnification. Third, the higher the NA, the less the depth of focus. The depth of focus (DOF) is proportional to 1 divided by the square of the NA. Note that the depth of focus is different from the free working distance. The free working distance places a critical limitation on the depth that the objective can focus through the specimen before the specimen makes contact with the tip of the objective. If we require that the microscope objective will be able to focus through a cornea that is 500 μm thick in its central region, then the free working distance of the objective selected for this specimen must be greater than 500 μm.

In the epi-fluorescence mode, the microscope objective functions as both the condenser and the collector of the fluorescence and reflected light. In that case, corrections for axial and lateral chromatic aberrations are critical, not only in the microscope objective but in other optical components such as the tube lens. With epi-fluorescence, the microscope uses illumination from above and an epi-fluorescence filter block that directs the excitation light toward the microscope objective and excludes the excitation light from passing to the tube lens and the ocular.

The excitation light is separated from the fluorescence light with the use of a **dichroic mirror**, a special type of interference filter in a fluorescence microscope that uses epi-illumination. It is designed and constructed to reflect selectively the shorter-wavelength excitation light and transmit the longer-wavelength fluorescence light. Dichroic mirrors or filters contain multilayer (20–50) thin films. They work by reflecting unwanted light back to the light source.

The threads of a microscope objective have been standard RMS thread size since the mid-1800s. In order to meet the new requirements of microscope users who may require a high-NA immersion microscope objective with a low power, it was necessary for both Nikon and Leica to widen the threads of their objectives. This means that if the microscope is designed for RMS-threaded microscope objectives, it is not possible to use the wider-threaded, non-RMS microscope objectives. The new standard is a 25-mm thread diameter and a 60-mm parfocal shoulder height. Today, the user has a wide choice of excellent microscope objectives from several manufacturers. Their websites list the characteristics of each microscope objective and information on the proper selection, care, and cleaning procedures. These modern microscope objectives are highly corrected for many types of optical aberrations.

Examples of advances in optical microscopes and their objectives can be found in some recent Nikon designs, whose features are advantageous in epi-fluorescence and differential interference microscopy. Nikon microscopes have a tube lens focal length of 200 mm that has minimal chromatic aberrations and images the specimen plane onto the primary image plane. The ocular, which is also designed to have minimal chromatic aberrations, images the primary image plane onto the eye. Their microscope objectives, with minimal chromatic aberrations, have a parfocal distance of 60 mm, a large diameter (high NA), and use a thread size of 25 mm.

Both axial and lateral chromatic aberrations are corrected independently in the microscope objective and the tube lens. The use of a 200-mm tube lens has an advantage over shorter tube lengths in that it creates a smaller angle between the light rays passing between the optical axis (lens center) and those off-axis. This causes a minimal shift of the light rays in the image plane between the center of the field and its periphery, which results in a sharper image.

Image brightness is another very important parameter. In epi-illumination microscopes, for lenses of equal magnification, image brightness is proportional to the fourth power of the NA.

The final resolution and quality of the image in an optical microscope is a function of several components: the illumination system and light source, the microscope objective, and the other optical components such as the tube lens, beamsplitters, filters, and polarizers. The resolution and noise characteristics of the solid state detector as well as the SNR and the quantum nature of light will affect image quality. The use of a highly corrected microscope together with a tube lens that only partially corrects for optical aberration will not produce the maximum image quality.

Finally, there can be variations among a set of microscope objectives even with the same part number from the same manufacturer. It is best to obtain a trial set of the same type of microscope objective and test each one in the microscope under identical conditions in order to select the microscope objective that produces the best image. Even with the selection of a highly corrected microscope objective, the use of a cover glass of incorrect thickness (thicknesses differ within the same box) or the incorrect immersion fluid, as well as dust, dirt, and mechanical imperfections on the optical surfaces, will severely degrade image quality and lower the image resolution.

The modern optical microscope may have fiber optic components, and high-power pulsed or continuous-wave lasers. Many optical components can be damaged and destroyed by inappropriate use with high-power lasers. Similarly, if the laser beam is not accurately centered in an optical fiber, it is possible to destroy the cladding.

The modern light microscope has brought about technical solutions to many earlier problems: stable mechanical structures have been developed, many types of optical aberrations have been corrected and minimized, and fluorescence microscopes have contributed to many advances in biology and medicine.

2.7 Sets of Conjugate Planes in the Optical Microscope

Conjugate points are those in both object and image space that are imaged one on the other. **Conjugate planes** are those perpendicular to the optical axis at the conjugate points. The **optical axis** is a straight line joining the centers of curvature of lens surfaces.

The light microscope, when analyzed by geometrical optics, consists of two sets of conjugate planes (see Fig. 2.2). Conjugate planes are imaged into each other; therefore, they can be viewed simultaneously. An object located at one conjugate plane will be imaged at each subsequent plane of that series. There is a group of four field planes, and a second group of four conjugate aperture planes. When using Köhler illumination, there are the following conjugate **aperture planes**: the light source (the lamp filament), the front aperture or focal plane of the condenser (located where the condenser iris diaphragm is situated), the rear or back focal plane of the microscope objective (back aperture of objective lens), and the plane of the eye point (coincident with the pupil of the eye) in the back focal plane of the eyepiece (exit pupil of eyepiece).

The set of conjugate object, image, or **field planes** consists of the field diaphragm of the lamp, the plane of the specimen, the real intermediate image plane, and the image on the retina, the film, or the faceplate of a video or solid state detector. Field planes are conjugate with the focused specimen. When the compound light microscope is correctly adjusted, the sharp image of the field stop diaphragm can be seen at the edge of the microscopic field.

If the compound light microscope is correctly adjusted, then these two sets of conjugate planes are independent of each other. For example, it is possible to close the condenser diaphragm to reduce the angular aperture of the illumination, but it will not change the area of the specimen that is illuminated.

The concepts of independent sets of conjugate planes are useful to understand both Köhler and critical illumination. As previously stated, Köhler illumination provides even illumination of the specimen because the filament (often not a uniform illumination source) is not imaged into the specimen plane (see Fig. 2.3), but rather the uniformly illuminated field diaphragm and the immediate adjacent lamp lens. The older method of critical illumination produced a brighter although uneven illumination, and the lamp filament is observed superimposed on the specimen.

2.8 Epi-Illumination Fluorescence Microscope

In a transmission optical microscope (see Fig. 2.1), the illumination light is on one side of the specimen, and the light from the specimen used in image formation is from the opposite side. When the sample is opaque, transmission light microscopy is not possible and other microscope configurations are employed.

Several terms exist for optical microscopes in which the illumination light and the light from the specimen are limited to the same side of the specimen: reflected light, incident light, epi-illumination, or metallurgical microscope. For example, in the metallurgical microscope, highly polished metal samples are observed with illumination and observation occurring on the same side of the sample. An incident-light microscope is one in which the microscope objective serves as its own condenser.

The term **reflected-light fluorescence illuminator**, or **vertical illuminator**, refers to the fact that the incident light is perpendicular to the optic axis; it is reflected toward the microscope objective by a dichroic mirror placed at 45 deg. to the optical axis (see Fig. 2.4). The same dichroic mirror excludes the excitation light, but passes the longer-wavelength fluorescence from the specimen. This is generally called **episcopic** illumination, referring to an illumination system situated above the microscope objective and including epi- and vertical illumination.

Both the fluorescence confocal and the multiphoton excitation microscope can be used as incident light microscopes. When incident light comes from above the microscope objective, there are special design considerations. First, for fluorescent incident light microscopy, the objective acts as both the condenser and the objective (see Fig. 2.4). In order to avoid reducing the NA of the objective, which would result in a reduction of the resolution and image brightness, the aperture iris is not located in the pupil of the objective.

Instead, the aperture iris is alternatively located between the light source and the field iris, and a lens images the aperture iris onto the pupil of the objective (see Fig. 2.4). Again, the objective also serves as an illumination condenser; however, in this case the aperture iris is imaged onto the back focal plane of the microscope objective. This design does not block the image-forming light path of the microscope.

A filter cube is used to direct the incident light towards the microscope objective, which focuses it on the object, and to separate the fluorescence from the incident excitation light. The filter cube contains three components: the exciter filter, the dichroic mirror, and the barrier filter. The exciter filter is a bandpass filter that limits the bandwidth of the excitation light if the source is broadband, such as a xenon arc lamp. The dichroic mirror reflects the excitation light towards the objective and the specimen, and serves to separate the excitation from the fluorescence light since the fluorescence light occurs at a longer wavelength. It passes the fluorescence light towards the tube lens and eyepiece for detection. Both the barrier filter and the dichroic mirror prevent the incident light and scattered light from the specimen from reaching the detector.

There are some limitations to the epi-fluorescence microscope. Important considerations in the ultimate sensitivity of the fluorescence microscope include SNRs and the quantum nature of light, as well as the nonspecific background light. In the ideal case, the only light detected is from the fluorescent molecules in the specimen. In practice, both the microscope and its components as well as the specimen contribute to background light.

The filters and dichroic mirrors used in the epi-fluorescence microscope are not perfect. Grease and dirt can also contribute to background fluorescence, as well as the specimen, the immersion oil, the slide, and the cover glass. Nonspecific autofluorescence at the wavelength of excitation and emission that corresponds to the absorption and emission properties of the specific fluorescent probe used in the measurement also limits sensitivity. Photobleaching of the fluorescent probe during the course of microscopic observation will reduce the signal; and if nonspecific components are less sensitive to photobleaching, the ultimate sensitivity is reduced. Single-molecule fluorescence detection is within the capacity of the fluorescent microscope; if the conditions are appropriate, extremely high sensitivity is achievable.

Problems not yet discussed include *in vivo* microscopy; light microscopy of thick, highly scattering specimens; and specificity and contrast in light microscopy. The next chapter presents the theory of image formation in a microscope and the seminal contributions of Abbe.

2.9 Summary

- A microscope provides a magnified image for the observer, although its most important function is to increase the resolution compared to that of the naked eye.
- Optical microscopy depends on both sufficient resolution and sufficient contrast.
- To understand Köhler illumination, it is useful to know the concepts of independent sets of conjugate planes.
- Köhler illumination provides several advantages: the field of view is homogeneously bright, the maximum obtainable resolution can be achieved under the given conditions, and the condenser aperture and size of the illuminated field can be varied independently.
- For incident light fluorescence microscopy, in which the objective also acts as the condenser, the aperture iris is located between the light source and the field iris, followed by a lens that images the aperture iris onto the objective pupil.
- Modern microscope objectives are manufactured to minimize the five categories of monochromatic Seidel optical aberrations: spherical aberrations, coma, astigmatism, field curvature, and distortion. Objectives are also corrected for chromatic aberrations.
- Many modern microscopes use infinity-corrected microscope objectives. An example, manufactured for fluorescence microscopy, is the Nikon Plan Apo 60× oil-immersion, NA 1.40, with a free working distance of 0.21 mm. This objective is designed to work optimally with a microscope that has a tube length of 200 mm, a thread diameter of 25 mm, and a parfocal distance of 60 mm.
- The use of a cover glass of incorrect thickness or the incorrect immersion fluid, as well as dust, dirt, and mechanical imperfections on the optical surfaces, will severely degrade image quality and resolution.
- There are variations within a set of the same type of microscope objectives. Test them all to select the best optical quality.

Chapter 3

Abbe Theory of Image Formation and Diffraction of Light in Transmitted Light Microscopes

The late 1800s saw a great effort on the part of microscope manufacturers to understand the basis of image quality, resolution, and contrast in their products. There was also a perceived commercial advantage in having a product that could be marketed as "scientifically designed." Both of these aims were achieved through the work of Ernst Abbe, a physicist who worked at the microscope factory of Carl Zeiss.

This chapter explains the seminal contributions of Abbe to the construction of microscope objectives and his diffraction theory and its role in image formation in the light microscope. The diffraction theory of Abbe and his experimental sets of gratings, apertures, and lenses used to observe the diffraction pattern in the back focal plane of the microscope objective provided evidence that there is an upper limit to the ability of a lens to resolve very fine spatial details.

An optical imaging system uses surfaces that refract (lenses) and/or reflect (mirrors) the light from an object to form its image. This chapter also explains how a lens forms the image of an object; describes the role of diffraction in the process of image formation; and shows how the collection angle of the objective, the refractive index of the medium between the specimen and the objective, and the wavelength of the light affect the limiting optical resolution. Finally, it discusses how optical aberrations confound this limit, derived on the basis of diffraction theory in the absence of aberrations.

3.1 The Contributions of Abbe

In the beginning of the 19th century, there were attempts to provide a scientific basis to imaging in the optical microscope. Both Fraunhofer and Airy attempted to use the theory of diffraction and interference to understand image formation. Nevertheless, it was in Jena, Germany, where Ernst Abbe made his important contribution to this problem.

Abbe studied physics and mathematics, first two years at University of Jena, and then three additional years at University of Göttingen, where he concentrated on the theory and practice of precision measurements. Abbe took courses that included individual practice on the construction of precision measuring instruments. When Abbe selected a topic for his Habilitation work, which would permit him to teach in a German university, he chose to study the mathematical analysis of precision and experimental error.

In 1866, Carl Zeiss made a request to Abbe: Could he place the construction of optical microscopes on a firm scientific basis? After many experiments, Abbe gained a deeper understanding of the principles of image formation in the optical microscope. He made two theoretical breakthroughs: first, that the influence of angular aperture on microscope resolution is a result of light diffraction caused by the specimens; and second, the so-called Abbe sine condition. This section introduces and describes these concepts, while the following section presents a detailed analysis of Abbe's diffraction theory.

Abbe also made several other major contributions to the field of microscopy; among his inventions are the first planachromatic and apochromatic objectives, lens designs based on his sine condition, and an interference test to determine lens curvature. It is of historical interest that in 1827, the English botanist Robert Brown discovered what was later termed "Brownian motion." This discovery occurred shortly after the discovery of achromatic microscope objectives. Improved optical instruments lead to discovery.

In geometrical optics, we often see the term **paraxial theory** (or **paraxial formula**), which applies to light rays that are sufficiently close to the optical axis so that $\sin \Theta = \Theta$, where Θ is the angle that a focused or divergent ray makes with the optical axis.

The foregoing approximation is known as the **paraxial limit,** also called the **Gaussian approximation**. Within this limit, the law of refraction can be linearized and the effect of a lens on light rays can be expressed in two simple statements. First, a lens has a focal length on the front and on the back side. These two focal lengths are defined on a common axis through the center of the lens. Second, light rays parallel to the common axis exit the lens on a path that intersects the focal point on the opposite side. A ray from the object that passes near the focal point emerges from the lens parallel to the optical axis.

A **lens** is a refractive element that affects the path of light rays. In this book, all lenses are composed of surfaces that are rotationally symmetric about the optical axis. Light rays from a point in the **object plane** travel towards the lens; the entrance pupil or aperture limits which rays reach the lens. The **exit pupil** or aperture limits which rays leave the lens and are converged or brought into focus at a point in the **image plane**. Even in the absence of physical apertures in front of and behind a lens, there are still entrance and exit pupils that result from the finite lens size.

Paraxial theory is not sufficient to calculate microscope resolution limits or design optical systems that minimize optical aberrations in microscope objectives. The theory of optical image formation and the derivation of the resolution limits of a lens depend on the use of physical optics, which incorporate the wave properties of light and explain such phenomena as light diffraction. Modern microscope objectives, with their diffraction-limited resolution, are designed in conformity with the Abbe sine condition, which is a testament to Abbe's seminal contributions.

The Abbe sine condition has important practical consequences that were valid historically and are equally valid today. It requires that optical aberrations in a microscope will be minimized when the microscope is used in conformity with its design conditions. In practical terms, this means that a microscope objective is designed to be used with a specific tube length. Abbe used his sine condition to design aplanatic microscope objectives that had minimal spherical aberration and coma. These microscope objectives contributed to the advances in biology and medicine described in the previous chapter.

The **Abbe sine condition** is an exact ray-tracing expression to calculate the positions of all rays that enter a lens. Historically, Abbe reasoned that there must be a relationship between the direction of rays (angle or ray with respect to optical axis) on the object side and that on the image side of the lens. Abbe found that for conjugate points, the ratio of these two direction angles must be constant over the full aperture of the optical system.

For the case of object and image planes oriented perpendicular to the optical axis, given a lens with transverse or lateral magnification, M, let U be defined as the angle relative to optical axis of a ray from an axial object point. U can also be defined as the object-side angle between a marginal ray and the optical axis; U' is the image-side angle between a marginal ray and the optical axis. The Abbe sine condition maps the angle U of a ray entering the objective to the angle U' of the ray as it arrives at the image plane. The Abbe sine condition is then given as $n \sin U = M n' \sin U'$, where n and n' are the refractive indexes of the medium on each side (object and image space, respectively) of the lens.

This means that optical aberrations in an optical microscope will be minimized when it is used in a manner consistent with its design conditions. With these two principles, Abbe designed aplanatic microscope objectives that have minimal spherical aberration and coma.

Abbe showed that the **resolving power** of a microscope, i.e., the minimum distance between two points that can be resolved (separated), is a function of the wavelength of the light used to form the image and the angular aperture or cone of the microscope objective (2Θ). The symbol Θ is half the angular aperture of the microscope objective, and n is the refractive index of the medium in the space between the objective and the specimen. With these, Abbe defined the numerical aperture ($NA = n \sin \Theta$). Furthermore, Abbe designed the microscope objectives and the compensating eyepieces to work together to correct residual aberration.

The NA is a critical feature of the microscope objective, indicating the light acceptance angle. This angle determines both the resolving power of the objective and its depth of field.

By 1889, Abbe had designed both water- and oil-immersion microscope objectives. His immersion objective, designed for monobromonaphthalene immersion, had a NA of 1.60! Another important advance was his substage illuminator. This device had an adjustable iris and permitted easy adjustment of oblique illumination. Abbe also contributed the first planachromatic and apochromatic objectives, a lens designs based on his sine condition, and an interference test to determine lens curvature.

There is another practical note for modern microscope users: in order to achieve the theoretical resolution limit for specific conditions, it is necessary to use the full NA of the microscope objective. In practical terms for fluorescence microscopy, this means that the projected diameter of the light source should match the diameter of the back pupil of the microscope objective. Alternatively, for laser-scanning microscopes, the Gaussian beam diameter should match the back pupil of the microscope objective. Failure to follow these conditions will result in a loss of NA and consequently a loss in resolution.

3.2 Abbe Diffraction Theory of Image Formation and Optical Resolution in the Light Microscope

A great advance in the understanding of light microscopy occurred with the seminal work of Ernst Abbe (1873) in Jena, Germany, on the analysis of image formation and resolution of a lens based on wave diffraction theory. Geometrical optics could not provide an adequate foundation for the analysis and understanding of the phenomenon.

Before discussing Abbe's diffraction theory, it is necessary to present several definitions. There are two extreme cases for image formation in a microscope: incoherent and coherent illumination. Examples of the former case include self-luminous objects such as a light bulb filament. Abbe worked out his theory based on coherent illumination. An example is a thin object illuminated by a small source with a low-aperture microscope condenser.

Optical path length or **distance**, for a homogeneous medium, is defined as the product of the geometrical length of the optical path and the refractive index of the medium in that path. **Optical path length difference** is the difference in length between two optical paths resulting from differences in geometrical length, refractive distance, or both. The term refers to the difference between two coherent wave trains that may interfere.

Diffraction is the deviation of the direction of propagation of light or other wave motion when the wavefront passes the edge of an obstacle. A **diffraction pattern** is formed in the back focal plane of the microscope objective, and is a distribution of intensities varying with direction in a regular manner and resulting from interference between portions of the diffracted radiation having differing phase relationships. **Diffraction limit of resolving power** or **diffraction-limited resolution** is a fundamental limitation imposed upon the resolving power of an optical system by diffraction alone, not by aberration.

Interference is the mutual interaction between two or more coherent wave trains. Interference converts optical path length differences in the object into intensity variations in the image and thus provides contrast. A **conoscopic image,** the interference pattern and diffraction image observed at the back aperture of the objective lens, is a two-dimensional projection of the rays traveling in three dimensions in the specimen space.

The **Airy pattern** is the intensity response for an aberration-free lens that has a circular pupil or aperture in front of it. The diffraction pattern of a point source that appears in three-dimensional space in and near the focal plane is called the **three-dimensional diffraction pattern**. In the presence of an aberration-free, diffraction-limited optical system, the two-dimensional slice of the diffraction pattern in the focal plane is the **Airy disk** and its surrounding diffraction rings. Above and below the focal plane, the Airy disk pattern on the axis of the light beam changes periodically so that the bright and dark Airy disk patterns appear alternately.

How does the size of the Airy disk vary as a function of the wavelength of the illumination light and the NA of the microscope objective? For the image of a diffraction-limited spot by a microscope objective, the following formula holds:

$$r = \frac{1.22\lambda}{2\text{NA}}, \tag{3.1}$$

where r is the radius of the spot for a self-luminous point in the image plane (i.e., a subresolution fluorescent bead), λ is the wavelength of the illumination light, and NA is the numerical aperture of the microscope objective. The formula indicates that the radius of the Airy disk will increase as the wavelength increases, and decrease as the NA increases. Therefore, the size of the Airy disk will become smaller with shorter wavelengths. As the NA increases, the Airy disk becomes smaller; the Airy disk from a point imaged with a microscope objective that has an NA of 1.0 is smaller than the Airy disk image from a similar point imaged with a microscope objective that has an NA of 0.4.

Now that we have explained some of the concepts and defined the key terms, we present the Abbe diffraction theory of image formation in the light microscope. Abbe observed, using a diffraction grating, that the diffracted light from a periodic specimen produces a diffraction pattern in the back focal (diffraction) plane of the objective lens.

First, I present a summary of the Abbe theory, and then some additional details of his theory of image formation in the light microscope. Abbe proposed that there is interference between the zero order and higher order diffracted rays from the specimen, and that this interference produces contrast in the image and determines the maximum spatial resolution with a specific microscope objective (for a fixed NA and wavelength of the illumination). The zero order is the undeviated or nondiffracted light. Abbe then went on to propose that at least two different orders of diffracted light must enter the objective lens in order to have interference in the image plane. To repeat, the Abbe theory has three parts: (1) the specimen diffracts the illumination light, (2) the diffracted light enters the microscope objective, and (3) interference of the diffracted and nondiffracted light occurs in the image plane.

According to Abbe's theory, light from a plane wave is incident on a grating-like object. The light is diffracted and forms a diffraction pattern in the back focal plane of the microscope objective. Every point in the back focal plane can be considered a source of coherent secondary disturbance. The light waves from all these secondary sources interfere with each other and form the image in the image plane of the objective (see Fig. 3.1).

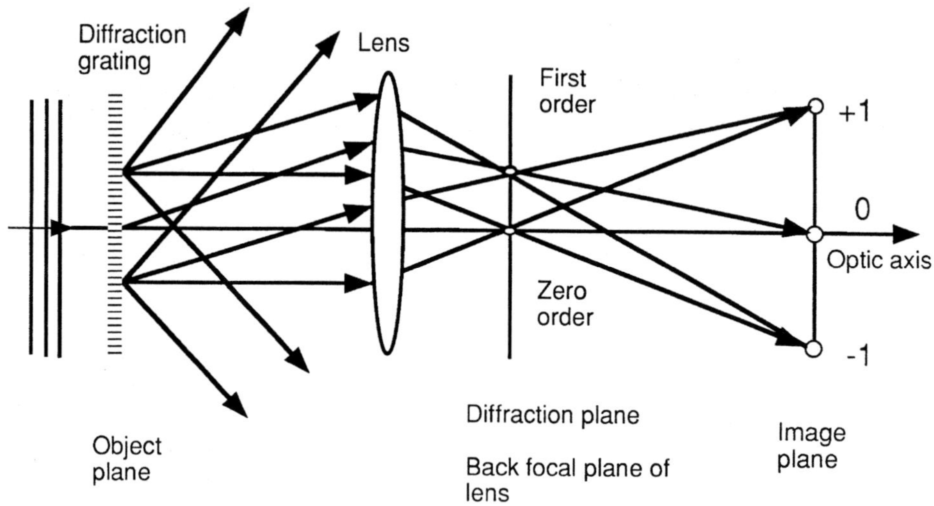

Figure 3.1 Schematic diagram to illustrate Abbe's theory of image formation.

Abbe assumed that the object is an amplitude grating; that is, a grating made of alternating opaque and transparent strips. The lens must have a sufficient aperture to transmit the entire diffraction pattern produced by the object; if the aperture only transmits a portion of the diffraction pattern, the resulting image will resemble a virtual image that would result in a diffraction pattern that the aperture transmitted. (See Fig. 3.1) If the spatial frequencies of the object are so high and the aperture so narrow that no portion of the diffraction pattern from the fine details of the object are transmitted by the aperture of the lens, then that detail will be invisible no matter what magnification is used!

Abbe demonstrated these concepts and conclusions with sets of diffraction gratings and apertures that transmit various orders of the diffraction pattern. He was not the first to observe the image in the back focal plane. Giovanni Battista Amici (1786–1863) first routinely used a lens to do so. Even so, it was Abbe who showed that the lens forms an image of the diffraction grating by the process of combining in the focal plane the harmonic components of the diffracted light. When the lens aperture is made increasingly larger, higher orders of diffracted light from the object are combined, and the image of the object contains increasingly more detail of higher spatial frequencies and thus is a closer representation of the object. Subsequently, Porter in 1906 used Fourier's techniques to develop the mathematical foundation of Abbe's theory.

What does the Abbe theory mean for the user of the optical microscope? To obtain the maximum resolution: (1) a microscope objective with a high NA should be used in order to accept more of the higher orders of diffracted light; (2) the illumination must be the correct NA to correspond to the NA of the microscope objective; (3) the shortest wavelength of light consistent with the objective corrections should be used, and (4) in transmitted light microscopy, the adjustment of the NA of the condenser is critical since it provides illumination over a wide range of angles.

What is the central zero-order diffraction spot in the image? This spot corresponds to the incident light that passed undisturbed through the specimen. Because the light rays in this spot are not diffracted, they contribute to the even illumination of the image plane. The adjustment of the condenser aperture is critical in setting the effective NA of the objective in the transillumination mode. The condenser diaphragm limits the angle of the illumination cone of light that reaches the objective and therefore limits its effective NA. If the condenser diaphragm is incorrectly adjusted, the illumination cone of light will reduce the NA of the objective and the back aperture stop of the objective will not be completely filled. This is illustrated in a transmission light microscope, where the adjustment of the condenser aperture has a direct effect on the spatial resolution of the microscope.

The bright spots in the image plane correspond to constructive interference of the light; the dark spaces between them correspond to destructive interference. The nondiffracted zero-order rays and the first- and higher-order rays are spatially separated in the back focal or diffraction plane of the microscope objective, but they are combined in the image plane.

Abbe experimentally demonstrated that it is a necessary condition that at least two different orders of light must enter the microscope objective in order for interference to take place in the image plane. If the specimen is composed of a periodic set of lines with interline spacing d, then a similar periodic pattern with spacing D will be formed in the diffraction plane, where D is inversely proportional to d.

To summarize, the minimum requirement for the resolution of a given periodic spacing of lines in an object is that at least the first-order light diffracted from the spacings of the object and the zero-order must be collected by the lens aperture. For coherent radiation, the minimum resolvable spacing is given by the wavelength divided by the refractive index of the media multiplied by the sine of the aperture angle of the objective.

At this point, it is useful to explain the coherence properties of light. The resolution of a microscope is a function of the design and construction of the optical elements, the presence and partial correction of various types of optical aberration, and the wavelength of the light used for illumination. In addition, the coherence properties of the light also affect its distribution in the image and resolution. Incoherent illumination implies that there is no definite phase relation between the intensities from adjacent points on the specimen. The intensities from each point of the specimen just add together in the image. With coherent illumination, the amplitudes from each point add up, but the final distribution is a function of their phases. Note that coherent and incoherent are two extremes; however, partially coherent illumination is closer to physical reality.

Some points on the basics of coherence theory are necessary. Usually coherence effects are separated into two classes: temporal and spatial coherence. **Temporal coherence** is related to the limited bandwidth of the light source; **spatial coherence** to the finite size of the light source.

In summary, what are the consequences of the Abbe diffraction theory? In order to have image formation in the image plane, the object must diffract the incident light, and the microscope objective must capture the diffracted light. In order to have image formation, portions of two adjacent diffraction orders must be captured by the microscope objective, which produces a barely resolved image. A sharply defined image of high resolution requires multiple orders of diffracted light to enter the NA of the microscope objective. Also, the sample must be illuminated with partially coherent light and there should be a coherence relation between the zero order and the diffracted rays that interfere to form the image.

The interference in the image plane results in image contrast. At least two different orders of diffracted rays must enter the lens for interference to occur in the image plane. The coherent light beams coming from the various parts of the diffraction pattern mutually interfere and produce the image in the front focal plane of the eyepiece. A result of Abbe's analysis is the understood importance of the use of high-NA microscope objectives.

The result that Abbe derived is that the limit of resolution is given by a numerical factor multiplied by the wavelength of the light divided by the NA of the objective. The numerical factor is a function of the form of the object and the aperture. With a slightly different numerical factor (which is somewhat arbitrary), the Abbe resolution limit is similar for both coherent and noncoherent illumination. The most important point of Abbe's theory is that for light of a given wavelength, the resolving power of a microscope objective is determined by the NA of the object.

The **diffraction-limited resolution** of a conventional light microscope is given by the **Abbe equation**. Abbe's diffraction theory placed a limit on the spatial features that can be resolved by a light microscope using oblique illumination. Abbe derived several equations for the maximum resolution under a variety of conditions. The smallest intensity detail that can be resolved with a microscope under direct illumination, and with a non-immersion objective, as a function of wavelength measured in a vacuum, λ, and the NA of the microscope objective is

$$\Delta x \equiv \frac{\lambda}{\mathrm{NA}}. \tag{3.2}$$

For the case of **oblique coherent illumination**, a parallel beam of light is inclined to the optical axis in such a manner that the direct light from the specimen just enters the microscope objective (see Fig. 3.2). Under these conditions, the microscope objective can capture the first-order diffracted and the zero-order nondiffracted beam. The first-order diffracted beam of the other side of the cone is lost and cannot enter the microscope objective. In contrast, the zero-order diffracted beam and one of the first-order diffracted beams can interfere and form a diffraction pattern in the image plane. Under these conditions, the maximum resolving power of the amplitude object (the periodic grating) is doubled, i.e.,

$$\Delta x \equiv \frac{\lambda}{2\mathrm{NA}}. \tag{3.3}$$

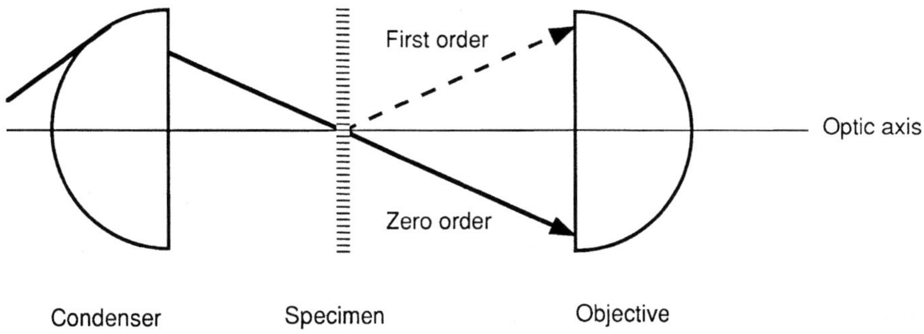

Figure 3.2 Schematic diagram showing oblique illumination. The specimen diffracts the light.

This result is based on the Abbe analysis of an object with amplitude that varied sinusoidally in space; he suggested that the light from the object could be considered as the superposition of two plane waves that move toward the lens and are inclined at an angle, Θ, to the optical axis. The object must diffract light, and this diffracted light must enter the lens in order for image formation to occur. If the microscope objective is not able to collect the plane waves, then they cannot contribute to the image formation. Thus, the resolution of the microscope is limited by both the wavelength of the illumination light and the NA of the microscope objective. The above relation is the diffraction-limited resolution of the microscope.

The experimental verification of the theoretical wave analysis of microscopic image formation was shown by Abbe. He used a diffraction grating for the specimen and observed its image in the microscope when the condensed aperture was closed down. Abbe demonstrated that there is a reciprocal relationship between the line spacing of the grating and the separation of the diffraction spots at the aperture plane. He observed the diffraction pattern of the grating on the image of the condenser iris diffracted by the periodic spacing of the grating. Each diffracted-order ray, including the zero-order ray, is focused in the back focal plane of the objective lens.

Abbe's most important experimental finding was that when the first-order pattern was blocked at the back aperture of the objective, the zero- and second-order patterns were transmitted. He found that the **orthoscopic image,** i.e., the image normally observed in a light microscope at the intermediate image plane and its conjugate planes, appeared with twice the spatial frequency due to the interference between zero- and second-order diffraction patterns. This remarkable result proved that the waves that form the diffraction pattern at the aperture plane converge and interfere with each other and form the image in the back focal plane of the objective. Abbe was able to further demonstrate that for the image of the diffraction grating to be resolved, at least the zero-order and the first-order diffraction patterns must be accepted by the NA of the objective lens.

Abbe diffraction theory explains the increase of resolution with the use of immersion microscope objectives. In a water-immersion microscope objective, the refractive index of the cells and the medium are almost identical; therefore, the aberrations that increase with depth within the specimen are minimal. When the specimen is mounted in a high-refractive-index material, the use of oil-immersion fluid with its refractive index of about 1.5 and a suitable objective is used in which the medium, immersion fluid, microscope objective, and cover glass have similar refractive indexes. In this case, the higher orders of diffraction from the specimen are correctly refracted and enter the NA of the microscope objective so that they contribute to the increased sharpness of the image. When an air objective is used to view the same specimen, many of the higher orders of the diffracted rays cannot enter the NA of the objective and the specimen is imaged at a lower resolution, with a concomitant loss of sharpness.

Another approach is based on the Fourier approach to wave optics and leads to the same result as Eq. (3.3). The number of spatial frequencies that can enter the microscope objective limit the image resolution. Thus, the Fourier series representing the image is truncated because of the NA of the objective, and this limits the spatial resolution of the image. Therefore, there is an upper limit to the ability of an optical system to resolve the spatial features in an object.

There are two consequences of the Fourier approach to image formation in the optical microscope. First, each point of the focal plane is characteristic of the entire specimen. Second, there is an inverse relationship between the dimensions of the specimen and those in the diffraction pattern. Finally, it is important to repeat that limited resolution of a microscope objective is a consequence of the fact that the finite aperture of the lens is unable to collect all of the diffracted light that leaves the specimen.

In conclusion, the Abbe theory of image formation in a light microscope set a resolution limit of approximately 180 nm in the focal plane and 500 nm along the optical axis. In recent years the Abbe limit on the optical resolution has been broken. These new far-field light microscopes are discussed in Chapter 13.

In the next chapter I further define various criteria for resolution and then describe how to characterize the performance of an optical microscope.

3.3 Summary

- Abbe made two theoretical breakthroughs: first, the influence of angular aperture is a result of the diffraction of light caused by the specimens, and second, the so-called Abbe sine condition.
- Abbe defined the numerical aperture as NA = $n \sin \Theta$.
- The smallest detail that can be resolved with a microscope as a function of wavelength, λ, and numerical aperture (NA) of the microscope objective is

$$\Delta x \equiv \frac{\lambda}{2\mathrm{NA}}.$$

- The first conclusion from the Abbe theory is that the resolving power of an optical light microscope is determined by the numerical aperture, not the magnification.
- The second conclusion from the Abbe theory is that shorter wavelengths will increase the resolving power of an optical light microscope.
- The Abbe theory yields a limit for far-field spatial resolution of the light microscope. The lateral resolution is approximately 180 nm, the axial resolution approximately 500 nm. This is valid for ultraviolet light and a NA of 1.6.

Chapter 4

Optical Resolution and Resolving Power: What It Is, How to Measure It, and What Limits It

This chapter defines and discusses criteria for optical resolution and indicates those factors that reduce it from its theoretical limits. Many of the definitions of resolution were originally derived for two-point resolution in telescopes used in astronomy and adopted for optical microscopy. Stars observed in an optical telescope can be considered to be point sources of light. Biological specimens observed in a light microscope are not point sources of light; however, the detection of fluorescence from single molecules is within the capability of a modern fluorescence microscope. Single fluorescent beads of submicron diameter can approximate point sources of light and therefore be used to experimentally determine the optical performance or resolving power of a microscope objective.

4.1 Criteria for Two-Point Resolution

A discussion of the criteria for two-point resolution begins with the definition of two related terms. **Resolving power** denotes the smallest detail that a microscope can resolve when imaging a specimen; it is a function of the design of the instrument and the properties of the light used in image formation. **Resolution** indicates the level of detail actually observed in the specimen. It depends on the resolving power of the microscope, the contrast generated in the microscope, the contrast in the specimen, and the noise in the detector.

Abbe's theory yields a limited far-field spatial resolution for the light microscope. The lateral resolution is approximately 180 nm, and the axial resolution approximately 500 nm. The Abbe theory of the role of diffraction and interference in image formation in the optical microscope leads to this summary of several important points and their consequences: First, the resolving power of a microscope objective is measured by its ability to differentiate two points. The smaller the distance between the two points that can be distinguished, the higher the resolving power. Second, as the wavelength of the light used to illuminate the source is decreased (shorter-wavelength illumination), two points can be resolved at a smaller distance of separation. Third, as we increase the NA, then two points closer together can be resolved.

Abbe's theory explains this in terms of higher orders of diffracted light from the specimen entering the collection angle of the objective. In terms of Fourier theory, this corresponds to higher spatial frequencies being imaged. With visible light, the minimal resolved distance between two points is of the order of 0.25 μm.

Several criteria for two-point resolution are based on theoretical cases with the absence of **noise**. Noise, which has the effect of reducing image contrast, may be expressed as the variation of a signal during repeated observation. Since noise is proportional to the square root of the average signal, if the intensity could be increased without saturation of the fluorescence and without photodamage to the specimen, then the signal will be increased. Alternatively, it can be integrated over time; since the signal is proportional to time and the noise is proportional to the square root of the integration time, the **signal-to-noise ratio** (SNR) will therefore be proportional to the square root of the time. One source of noise is the detector. Any detector used with an optical microscope, whether it is a camera, a solid state imaging detector, or the human eye, measures the **intensity** of the light, which is defined as the square of the amplitude of the electromagnetic field.

The **Sparrow criterion** and the **Rayleigh criterion** are used to define the resolution of an optical system that can resolve two points at a minimal distance of separation. The Sparrow definition is that two points of equal brightness are imaged as two separate points if the intensity at the midpoint between them is equal to the intensity at the points. The Sparrow minimal resolved distance is shown in Eq. (4.1). This is the relation for an incoherent imaging system, for example, two stars observed through a telescope. In the image plane of the telescope, the intensity at any point is equal to the sum of the intensities from each of the stars. If there is coherent imaging of two in-phase points, then the Sparrow definition is approximately 1.5 times larger than that for the incoherent definition.

$$\Delta x(\text{Sparrow}) \equiv \frac{0.51\lambda}{\text{NA}}. \tag{4.1}$$

Before discussing the Rayleigh definition, it is necessary to define the **Airy disk**, which was first described by Airy in 1828 in connection with astronomical telescopes: the image of a point object (zero extension in space as compared with the wavelength of light) that is imaged by an aberration-free lens with a finite aperture. The **Airy diffraction pattern for a circular aperture** is also called the **Airy pattern** or the two-dimensional **point spread pattern.** The Airy disk consists of a central peak of intensity surrounded by weaker intensity rings separated by dark rings. Approximately 80% of the incident intensity is in the central bright spot. The size of the central bright spot is proportional to the incident wavelength and inversely proportional to the NA.

A microscope forms an image of the object or specimen. The microscope is assumed to be aberration free and the image is formed solely by diffraction of the light by the specimen. Every point of the object is represented in the image not by a conjugate point, but by the Airy diffraction pattern. Therefore, the resolving power of the microscope objective can be determined by experimental measurement of the size of the Airy disk diffraction pattern, which is controlled by the wavelength of the light, the refractive index of the medium between the specimen and the microscope objective, and the NA of the objective and the microscope condenser lenses.

The alternative **Rayleigh definition** for the minimal separation of two points illuminated with incoherent illumination is as follows: two points of equal brightness can be imaged as two separate points if at the midpoint of their separation the intensity is reduced 26.5% from the intensity of each point. For incoherent illumination, this minimal resolvable distance for two separated points is 0.61 times the wavelength of illumination light divided by the NA. Incoherent illumination occurs with fluorescence or when the specimen is illuminated with a large cone of light from the condenser lens. There is no interference between adjacent Airy diffraction patterns, and therefore the intensity distribution pattern of two closely spaced or partially overlapping Airy diffraction patterns can be used as a criterion of resolution. The Sparrow criterion has the advantage over the Rayleigh criterion in that it is also applicable to coherent imaging.

The original definition that Rayleigh proposed in 1896 was that two points emitting incoherent light of equal intensity are resolved if they are sufficiently separated in space so that the center (maximum intensity) of the Airy disk of one point object is situated at a point that corresponds to the first minimum of the diffraction pattern of the second point object.

$$\Delta x(\text{Rayleigh}) \equiv \frac{0.61\lambda}{\text{NA}}. \tag{4.2}$$

It may be possible to detect unresolved objects. For example, colloidal gold particles can be detected but not resolved in a light microscope. Also, fluorescent labeled parts of the cellular cytoskeleton can be detected but not resolved. Objects of dimensions below the resolution limit can be detected if they have sufficient contrast against the background. However, their dimensions will appear to be that of the Airy diffraction pattern.

4.2 The Role of Depth Discrimination

Part II of this book introduces the confocal fluorescence light microscope, which is fundamentally different from the conventional or wide-field light microscope because it provides depth discrimination. With increasing defocus, the image does not become blurred; rather, it darkens and rapidly disappears. Depth discrimination is the key to optical sectioning; that is, the ability to acquire thin optical sections through a thick specimen and then to use a computer to reconstruct its three-dimensional structure.

Resolution is an important factor in a microscope. While it is a function of wavelength and NA of the objective, it also depends on noise, contrast, the exact nature of the specimen, and the type of illumination used. However, it is the advances in depth discrimination that have revolutionized the use of the light microscope, along with its ability to optically section a thick, living specimen.

We now discuss how to measure the axial or z-response of these types of microscopes, as well as how to measure the transverse resolution; that is, the resolution in the plane of the specimen.

One technique to obtain an experimental measure of the range or axial resolution is to measure the intensity of the light reflected from the source as a function of the distance from the focal plane of the microscope objective and a plane mirror placed on the microscope stage. This is performed by scanning a mirror axially through the focal plane of the microscope objective and measuring the intensity of the reflected light as a function of defocus distance. The **axial resolution** (along the optical axis) can then be defined as the width of the plot at the half-maximum intensity point. The presence of optical aberrations in the microscope results in asymmetry in the intensity versus distance plot and also creates sidelobes.

Another method to determine the optical resolution of the microscope objective is to obtain a **z-series** (a stack of optical sections through a specimen) of images of a microscopic spherical fluorescent bead smaller (0.1 μm) than the wavelength of light that acts as a point source of light. The composite stack of the images from above to below the bead will have finite height in the axial direction, and shape in the *x*-*y* dimension parallel to the specimen plane. There may also be asymmetries in the pattern as well as numerous sidelobes. The resolving power of a confocal microscope is lower in the axial direction (along the optic axis) than in the lateral (transverse) direction.

Finally, the **transverse resolution** (in the plane of the specimen) of the optical microscope can be estimated by imaging a standard microscope test specimen—a slide containing various patterns with different spatial frequencies or a biological specimen such as a diatom with independently measured, known distances between periodic lines. Another useful test object is an integrated microchip with known, independently measured line spacing.

4.3 Point Spread Functions Characterize Microscope Performance

In a diffraction-limited optical system, the microscope objective will form an image of a point object. The point object is assumed to be luminous, and the image is formed in the image plane. Because of the finite lens aperture, the image of a subresolution point will not be a point, but will be extended in three-dimensional space thanks to light diffraction. The quantitative description of this spreading is termed the **point spread function** (PSF). The optical PSF is related to the electrical circuit response to a delta function impulse (extremely narrow pulse); in fact, it is a two-dimensional optical analog.

There are two types of PSFs. The **amplitude PSF** of a microscope is related to the strength of the electromagnetic field in the image plane caused by a point source of light. The amplitude PSF is the transverse spatial variation of the amplitude of the image at the detector when a perfect point of light illuminates the lens. The squared modulus of the amplitude PSF is the **intensity PSF**. The intensity PSF is the spatial variation of the intensity of the image at the detector plane when a perfect point of light illuminates the lens.

In an ideal case, i.e., one that is free from all optical aberrations and in which the pupil of the lens is evenly illuminated, the PSF will be rotationally symmetrical and longer along the optical (z) axis than in the transverse (x-y) plane. If we observe the PSF in the image plane, we see the Airy disk pattern. In the x-z plane, the pattern is elliptical with numerous sidelobes. The PSF can be used to characterize the optical performance of a light microscope. In order to characterize both the resolution and the optical sectioning thickness, it is often useful to define the half-maximum of the PSF ellipsoid of rotation; that is, the area in which the three-dimensional PSF in both the axial and lateral direction has an intensity value of one half the central maximum.

The PSF extends in three dimensions; the Airy disk is composed only of those components in the x-y or lateral direction. The Airy disk is radially symmetric, with a central peak of intensity and sidelobes of decreasing intensity. Between the central peak and each successive sidelobe, the intensity is zero. The sidelobes are sets of concentric rings of decreasing intensity. Since the Airy pattern is formed from a circular microscope objective, the symmetrical Airy pattern represents the lateral intensity distribution as a function of distance from the optical axis.

In a conventional light microscope, the intensity PSF is the Airy disk, and its size gives the Abbe resolution in the focal plane. The lateral resolution in a conventional light microscope can be ideally described for incoherent illumination as the full width at half maximum of the central peak of the intensity PSF. The intensity PSF is at least 3 times longer in the axial direction as in the lateral or transverse direction. The PSF for the conventional microscope also has sidelobes of lower intensity. Finally, the axial response is constant; consequently, there is no depth resolution in a conventional microscope. (A number of techniques and approximations are used to calculate the theoretical PSFs. The detailed calculations using each set of approximations are given in the references [e.g., Corle and Kino] and are not discussed further in this textbook.)

For example, a comparison of various types of microscopes can be based on the lateral and axial extents and the intensity of the sidelobes relative to the central intensity peak. The lateral and axial extents of the intensity PSF determine the lateral and axial resolution of the light microscope. The lateral resolution of a light microscope is related to the size of the Airy pattern (the lateral components of the intensity PSF).

The topic of resolution in an imaging system, e.g., a light microscope, is very complicated. Typically, many assumptions are made in the analysis, e.g., aberration-free optical system, ideal lenses, point sources of light, infinitely small pinholes. Rarely do these ideal assumptions apply to an actual optical system. The resolution is a function of the type of object, i.e., a point, a line, a plane; it also depends on the type of illumination: is the light noncoherent, coherent, or partially coherent? A careful comparison of resolution requires that every assumption be clearly stated and all approximations be defined.

The lack of depth discrimination in the conventional light microscope represents its greatest limitation. This formidable limitation has numerous consequences. First, epi-fluorescence microscopy is limited to thin smears. Alternatively, extremely thin mechanical sections of a thick specimen can be observed with fluorescence microscopy. When thick fluorescent specimens are observed, the image is blurred and details in the focal plane are generally obscured, because fluorescent light from above and below the focal plane contributes to the image. Second, the lack of depth discrimination is also a problem in reflected light microscopy of thick, highly scattering specimens. In the past, thick specimens (for example, embryos) were fixed and mechanically sectioned with a microtome. The individual thin mechanical sections were then observed with a conventional light microscope. Finally, three-dimensional wax models were constructed to illustrate the structure of the entire embryo.

In addition to the multitude of problems and artifacts of fixation, the conventional microscope precluded the observation of dynamic events; for example, cell division, cell differentiation, fertilization, and many pathological processes. In neurobiology, developmental biology, and the clinical medical sciences of ophthalmology, the lack of depth discrimination was a barrier to progress.

In the absence of optical aberrations and noise, the phenomenon of light diffraction will limit the resolution of a light microscope. Real optical microscopes that have optical aberrations and noise will have a reduced resolution. The intensity PSF can be used as an indication of microscope performance. Alternatively, the transverse and axial resolutions can be determined. The next chapter describes techniques to provide contrast in the image.

4.4 Summary

- The analysis of resolution is confounded by many variables: contrast, noise, digitization, wavelength, type of object (point, line, plane), and degree of coherence of the light.
- Detectors measure the intensity of the light, which is the square of the amplitude of the electromagnetic field.
- The Sparrow criterion and the Rayleigh criterion are used to define the resolution of an optical system that can resolve two points at a minimal distance of separation.
- Although the pinhole is not an infinitely small point, the fluorescent laser-scanning confocal microscope has depth discrimination and suppresses stray light, which improves the image contrast.
- A wide-field (nonconfocal) light microscope has no depth discrimination.
- With the standard optical microscope, when the object is defocused, the image blurs.
- Three-dimensional light microscopy of living, thick specimens depends on the confocal microscope's depth discrimination, which is the basis of "optical sectioning."

Chapter 5

Techniques That Provide Contrast

From the theory of image formation developed by Ernst Abbe in 1872 until his death in 1905, there was a period of great advances in light microscopy. The use of apochromatic microscope objectives and oil-immersion objectives resulted in high-quality light microscopes that reached the theoretical limit of lateral resolution of approximately 200 nm (0.2 μm). The next important step was to improve image contrast. This chapter reviews several methods and techniques to provide contrast.

We define **contrast** as ratios of or differences of light intensities between different areas (or pixels) in the optical plane, such as a difference in intensity between different points of a specimen, or between a specimen and the background:

$$\text{Contrast} = \frac{I_{\max} - I_{\min}}{I_{\max} + I_{\min}}. \tag{5.1}$$

Contrast in an image is determined by several parameters: signal strength (number of detected photons), the dynamic range of the signal (lowest to highest level of signal), optical aberrations of the optical system, and the number of picture elements per unit area (**pixels**).

The technical advances that resulted in greatly improved contrast can be divided into two groups: nonoptical techniques, as exemplified by fluorescence microscopy and the development of epi-fluorescence microscopy; and optical techniques, such as phase contrast microscopy and differential interference microscopy. The techniques of fluorescence microscopy provide contrast as well as high specificity and sensitivity. Phase contrast and differential contrast microscopy permit the microscopic observation of live, unstained cells in tissue culture. Another important technique is video-enhanced contrast microscopy. These groups of techniques that provide and enhance image contrast have resulted in advances in cell biology, neurobiology, and developmental biology as well as diagnostic techniques in clinical medicine. Note that two types of microscopy can be combined, called **correlative microscopy**, to further minimize artifacts and the false interpretation of images.

5.1 Nonoptical Techniques

In the previous sections we discussed the use of stains and dyes as well as the great specificity that is possible with the use of immunocytochemical methods. The autofluorescence of organelles, cells, and tissues was known for a long time. Even so, the use of specific stains, dyes, and what we today call fluorescent probes or fluorochromes has truly revolutionized fluorescence microscopy. From the use of genetic fluorescent probes such as various types of fluorescent proteins that permit the location and monitoring of gene expression to *in situ* hybridization, these techniques are indispensable for studies of molecular and developmental biology.

Confocal microscopy can operate in both the reflected-light and the fluorescence mode. Multiphoton excitation microscopy operates only in the fluorescent mode and depends on either the autofluorescence of endogenous fluorescent molecules or the use of exogenous and genetic fluorescent probes. The great advantage of the fluorescent method is high specificity, and that advantage is exploited in many ingenious biological studies that use a variety of techniques: gene array chips, live cell and tissue studies, *in vivo* clinical microscopy of the brain, eye, and skin, and the use of cell sorters.

Fluorescence microscopy attains its remarkable specificity from the spectroscopic properties of the fluorescent molecules. These properties include the absorption and emission spectra, polarization and anisotropy of the fluorescence emission, rates of intersystem crossing, and the fluorescent emission lifetime. Other factors that affect fluorescence are the environment, quenching constants, and the proximity of nonradiative interaction with other molecules.

As previously stated, the eye and other light detectors are sensitive to light intensity. Fluorescence intensity is affected by several parameters in addition to the intrinsic fluorescence lifetime: concentration of the fluorescent molecule, the presence of quenchers, nonradiative energy transfer, the molecular environment, and the quantum efficiency of the fluorescence.

Measurement of the fluorescent lifetime of a molecule is much less sensitive to these quantities and therefore, in many cases, may provide additional information. Fluorescent lifetimes can be measured by either time-domain or frequency-domain techniques. The technique of **lifetime imaging microscopy**, in which the lifetime of the fluorescence forms the image and not the intensity as in standard fluorescence microscopy, is finding new applications in cell biology.

The clever measurement and analysis of one or more of these spectroscopic properties under various conditions has resulted in the development of many new, powerful, sensitive, and precise spectroscopic techniques with applications to cell and molecular biology, neuroscience, and clinical medicine. *In vivo* optical microscopy is an important technique for the detection and monitoring of disease progression. Modern developments include the imaging of cells and tissues based on a multiplicity of spectroscopic parameters; not only absorption and emission spectra, but polarization and fluorescence lifetime imaging show great promise for optical biopsy. To obtain a detailed presentation of these techniques, their theoretical foundations, and applications, the reader is referred to several books listed in the additional resources section.

The applications fluorescence resonance energy transfer (FRET), lifetime imaging (FLIM), and fluorescence recovery after photobleaching (FRAP) are rapidly expanding in cell biology. Another very sensitive fluorescence technique is fluorescence *in situ* hybridization (FISH), which is used to visualize and localize sequences of DNA and RNA on chromosomes, cells, and in tissues. It is based on hybridization between sequences of single-strand DNA of chromosomes or cell nuclei and fluorescent labeled complementary sequences.

Many of these techniques have been implemented in both fluorescence confocal microscopy and multiphoton excitation microscopy, for the study of molecular interactions, conformation changes, and signaling, protein and lipid trafficking in cells, *in vivo* measurement of molecular diffusion, gene expression, studies of cell-cycle regulation, developmental biology, cancer research, cell function and metabolism, cell death, monitoring intracellular ion concentrations, monitoring cell excitation, and the rapid detection of pathogens.

5.2 Optical Techniques

Previously, we mentioned the utility of both Zernike's development of phase contrast microscopy and Nomarski's DIC microscopy; the latter has in many cases surpassed phase contrast microscopy. This section expands the discussion of these two types of microscopes.

Some terms must be defined. An **interference microscope** is one based on **interference**, which is the mutual interaction between two or more coherent wavetrains. Interference is used to convert optical path differences in the object into intensity variations in the image, thereby forming contrast. Interferometer-based microscopes are designated by the specific type of the interferometer; for example, the Linnik interference and the Mach-Zehnder interference microscopes. **Differential interference** is a technique in which two wavetrains that fall on the object or the image plane are separated laterally by a distance similar to the minimum resolvable distance, creating double-beam interference. This type of contrast gives the impression of unilateral oblique illumination. In the transmitted-light mode, the variations in optical pathlength that do not result from differences in physical thickness appear as a relief in the image.

We discuss these two techniques; however, a number of optical methods are used to generate contrast in optical microscopy of live cells, tissue, and whole organisms, including various modes of interference contrast, oblique illumination, dark-field, single-sideband edge-enhancement, modulation contrast, polarization, Schlieren, and total-internal-reflection microscopy. These types of microscopy are described in the references.

5.2.1 Phase contrast microscopy

The use of light microscopy with living cells is difficult because the thin, transparent, unstained cells have little effect on the absorption of light in the bright-field light microscope. These specimens do affect the phase of the transmitted light, but not its intensity; the direct zero-order light travels through the specimen undeviated. The light deviated by the specimen is retarded by the specimen's thickness or refractive index. This scattered or deviated light arrives at the image plane out of phase with the direct light, but is similar in intensity. Since the eye and other light detectors are amplitude, not phase detectors, the changes in phase are not detected. Thus, the image of the specimen has no contrast. The problem is how to image unstained live cells in a light microscope.

The solution to this problem is of great significance. It permitted the microscopic observation of live, unstained cells in tissue culture and therefore the long-term observation and investigation of the dynamical processes that occur in cells and tissues. Frits Zernike, a Dutch physicist, solved this problem by designing and constructing a light microscope with phase contrast optics, which converted phase changes into intensity changes that could be detected. As a result of this invention, Zernike received the Nobel prize, which testifies to its significance.

Zernike had worked out the principles to convert differences in phase to differences in amplitude in the 1930s, but only received the Nobel prize in 1953. Zernike's Nobel prize address, "How I Discovered Phase Contrast," is available on the Web and is important to read. In his address, he related how an earlier technical method of Lord Rayleigh to produce phase stripes was critical to his invention. Zernike shortened the words "phase strip method for observing phase objects in good contrast" to phase contrast, thereby coining the term.

In 1935, Zernike published his first paper on phase contrast microscopy. The Zeiss Corporation in Jena constructed the first prototype of a phase contrast microscope in 1936. In 1941 Kurt Michel, head of the microscopy department at Zeiss Jena, made the first movie using a phase contrast microscope, which showed meiosis in the spermatogenesis of the grasshopper. Phase contrast microscopy then became a very important technique used to investigate the biology of live cells.

Figure 5.1 shows an optical schematic for the phase contrast microscope. The principle of the phase contrast microscope is the separation of the direct zero-order light from the diffracted light at the back focal plane of the microscope objective. These two light fields then interfere and form a high-contrast image based on intensity variations. The light passing the specimen is composed of two parts: (1) the plane waves that are present without the specimen, and (2) the light scattered as spherical waves by the refractive features in the specimen. **Refraction** is defined as a change in the direction of propagation of radiation caused by the change in the velocity of its propagation upon passing through an optically nonhomogeneous medium or upon passing from one medium to another in a direction other than the normal to the interface. Zernike knew that the scattered waves lag in phase by 90 deg. to the direct, unscattered light.

Zernike used two optical components to make the phase contrast microscope. The first is a phase annulus, placed in the back focal plane of the microscope condenser. The second is a phase plate placed in the back focal plane of the microscope objective. In the phase contrast microscope, the image from the condenser annulus must be correctly aligned with the groove (or ridge) on the phase plate in order for the nondiffracted light to have a phase advance or a phase delay.

How do these two optical elements work together to form the phase contrast image? With Köhler illumination, the phase or condenser annulus is an opaque

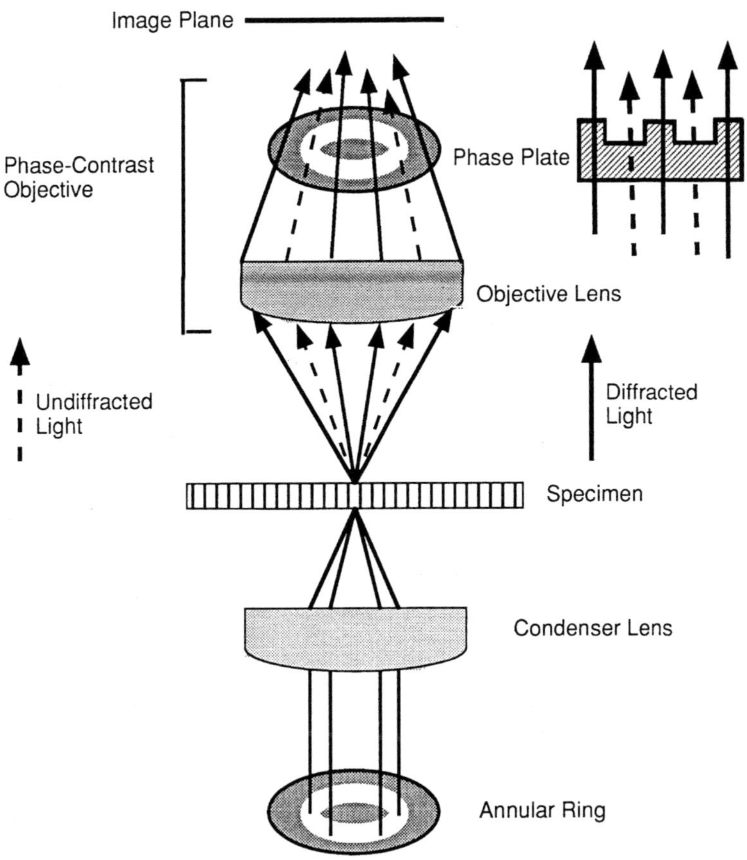

Figure 5.1 Schematic diagram showing the optical components of a phase contrast microscope. Diffracted light from the specimen is shown as solid arrows; undiffracted light is shown as dashed arrows.

black plate with a transparent annulus. When this is placed on the front aperture of the condenser, the specimen is illuminated by light beams that come through the transparent ring. The image of the phase annulus is formed in the microscope objective that is coincident with the phase plate built into the phase contrast microscope objective. The full amount of direct light is transmitted by the annular zone of the phase plate. At the same time, the scattered light is transmitted by the complementary zones of the phase plate.

The direct light transmitted by the annular zone in the phase plate undergoes two changes: it is attenuated to an amplitude similar to that of the scattered light, and the phase difference is shifted from 90 deg. to 180 deg. The phase plate is thinner in the attenuation zone that actually advances the phase of the direct light by 90 deg.

In the image plane of the phase contrast microscope, the direct light that is phase advanced is the reference beam of light over the full field of view. The scattered light amplitude field from the specimen interferes with the direct light to form the phase-contrast image. Since the two light fields have a 180-deg. phase difference, the interference results in a phase contrast image with high contrast.

The advantages of phase contrast microscopy include its simple design and use and its insensitivity to polarization and birefringence effects. The latter is a great advantage for viewing cells through plastic containers. There are also limitations: a special phase contrast microscope objective is required; the full illuminated NA is not used, which impairs the classical resolution limit; and the phase plate results in a "halo" artifact in the image when other modes of microscopy are used. Finally, care must be taken in interpreting phase contrast images, because the observed intensities do not necessarily correspond directly to object structures.

Phase contrast microscopy is very useful for the observation of living, unstained cells in tissue cultures well as slices. It can be combined with reflected light fluorescence microscopy to show areas of a specimen that are not fluorescent.

5.2.2 Differential interference contrast (DIC) microscopy

The DIC microscope is preferred over phase contrast microscopy for live cell and tissue imaging. Both the condenser and the microscope objective are used at the full NA, and Köhler illumination can be properly utilized; the result is improved axial resolution, the absence of halo artifacts, and images that can be processed to improve contrast.

The principle of DIC is that the phase shifts resulting from the refractive structure of the specimen are encoded in a field of polarized light. The two superposed components are both offset and analyzed to show refractive index gradients. The DIC microscope uses dual-beam interference optics that transform local gradients in optical path length in an object into regions of contrast in the image. Köhler illumination is required for correct location of the interference plane of the DIC prism in the conjugate aperture planes of the microscope condenser and the microscope objective.

How does DIC microscopy compare with phase contrast microscopy? In the phase contrast microscope, the intensities in the image are proportional to the difference in optical paths. In DIC microscopy, the images represent the rate of change of the optical path across the object in the direction of shear; in other words, the DIC image is proportional to the first derivative of the optical path difference. DIC images are formed from the rate of change of the optical path difference, instead of the absolute magnitude of the optical path difference that occurs in phase contrast microscopy. Therefore, DIC microscopy can image much thinner specimens at high contrast. DIC microscopy can also be used with white rather than monochromatic light, which results in color contrast that may add further contrast to the image.

The optical components of a DIC microscope include a linear polarizer located between the light source and the condenser, a modified Wollaston prism located close to the iris in the back focal plane of the condenser, a Nomarski prism behind

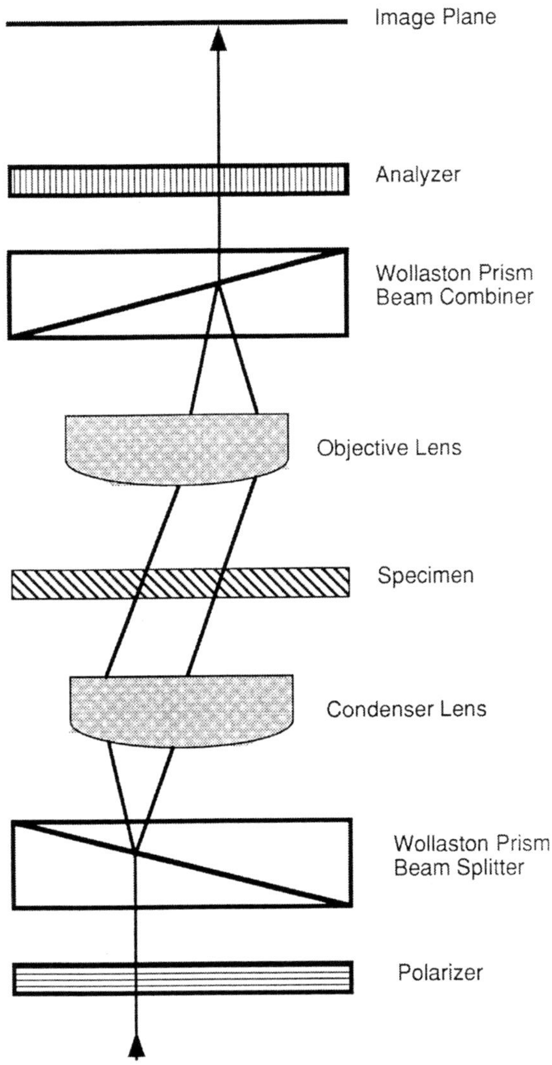

Figure 5.2 Schematic diagram illustrating the principle of a differential interference contrast microscope.

the microscope objective, and a linear polarizer or analyzer in front of the tube lens in the image plane (see Fig. 5.2). Following the light from the lower portion of Fig. 5.2 to the image plane at the top of the figure, the function of each component is explained. Light from the source (not shown) is passed through a polarizer; then the polarized light (which consists of light vibrating only in a single plane perpendicular to the light path) is passed through a modified Wollaston prism, or beamsplitter, which is located below the condenser lens. The Wollaston prism is made by cementing the two quartz crystal halves together. Quartz is a birefringent material, and incident light rays that have linear polarization are split or sheared into two components, each with perpendicular planes of polarization. **Birefringence** is the **double refraction** of light in certain materials (e.g., quartz, calcite) that are optically anisotropic due to the existence of orientation-dependent differences in refractive index. The incident ray is split or separated into two rays with different planes of polarization; they traverse different paths in the crystal and emerge as two rays that are linearly polarized. The electric field vectors of these two rays vibrate in perpendicular planes.

The plane-polarized light, which is vibrating in one plane perpendicular to the direction of propagation of the light, is split into two rays, each vibrating perpendicular to the other. These two rays travel in slightly different directions and intersect at the front focal plane of the condenser. Since these two beams vibrate perpendicular to each other, they cannot cause interference. These two rays leave the condenser, are parallel, and have a slight path difference. The distance between these two rays is denoted as the shear, and is less than the diameter of the Airy disk and the resolving power of the objective.

The two beams with perpendicular polarization and a small space between them (shear) pass through the specimen. The specimen's varying thickness, refractive indexes, and specifically the slope or rate of change of these quantities with distance (in the direction of shear) in the specimen affect the two beams.

The objective focuses both beams at the back focal plane. At that plane the two beams enter the beam-combing modified Wollaston prism, which removes the shear and the original path difference between the two beams. Since the beams passed through the specimen, parallel beams that passed through different regions of the specimen will have different optical path lengths. For the beams to interfere, the vibrations of beams of different optical path length must be brought into the same plane and axis. This function is performed by the analyzer placed above the beam combiner. When the source is white light, the rate of change of optical path differences within the specimen is observed in the eyepiece as differences in intensity and color. One side of a detail in the specimen appears bright or in one color, and the other side appears darker or in another color.

The normal Wollaston prism is constructed with the optical axis of each quartz wedge orthogonal to the other. If a normal Wollaston prism is placed between two crossed (perpendicular) polarizers, parallel interference fringes (alternating light and dark bands) are observed within the prism when it is viewed end on. The location of the interference fringes is called the interference plane, which is located within the normal Wollaston prism. Nomarski's modification was to construct the prism so that the optical axis of one of the wedges is oblique to the optical axis of the second wedge. Thus, the interference plane is displaced to a location outside the prism. The interference plane is now located several millimeters from the prism. This permits the interference plane to the beam-combining modified Wollaston prism to be within the back focal plane or diffraction plane of the microscope objective.

The upper beam-combining prism is movable and can be used to compensate for selected phase shifts within the specimen. The sensitivity of the microscope can be modified by sliding the combining prism. Note that individual Wollaston prisms are required for each microscope objective. The contrast based on light intensity and color is related to the rate of change in refractive index, thickness, or both in details or adjacent regions of the specimen.

DIC microscopy images the rate of change of the optical path difference, as compared with phase contrast microscopy, which images the absolute level of the optical path difference. Therefore, DIC microscopy can form very high-contrast images of much thinner specimens. Nevertheless, the technique has some limitations. Compared with phase contrast microscopy, DIC microscopy is expensive. Also, the use of plastic petri dishes, which are birefringent, will adversely affect the technique. Specimens that are very thin may be better observed using phase contrast microscopy. Apochromatic microscope objectives that affect polarized light are not suitable for DIC microscopy. And, as with phase contrast microscopy, the interpretation of the images can be difficult and misleading, and therefore caution is always advised.

DIC microscopy is one of the techniques that prevent out-of-focus light from contributing to the image. The out-of-focus refractive index changes are blurred and produce very weak gradients of optical path length difference in the plane of focus. Therefore, the light from outside the focal plane makes little contribution to the image.

5.2.3 Video-enhanced contrast microscopy

A **television microscope** is a microscope adapted so that its image is displayed on a television system. **Video-enhanced contrast microscopy** is another technique to enhance the contrast in a light microscope. It is a special form of television microscope in which the image is electronically processed in order to enhance contrast. Both Robert Allen and Shinya Inoué made independent developments of video-enhanced microscopy. This technique is valuable for the study of small features of low contrast such as those that occur in living cells.

In 1934, V. K. Zworykin developed the first microscope, which he called an electric microscope, to display an image by scanning. Zworykin built an ultraviolet microscope that included quartz optics in the microscope and the objective, an iconoscope TV camera to convert the image produced into a visible image, and a cathode-ray tube (CRT) as a display device. Zworykin was the first to demonstrate the control of the microscope magnification by varying the area that was scanned on the specimen.

Modern confocal microscopes, which are based on the Nipkow disk (see Sec. 7.3), can be made to show the image in real time and with true colors. Many of the modern confocal microscopes and multiphoton excitation microscopes display the image as gray level; that is, shades of gray are used to map intensity. The computer then assigns a color look-up table to convert gray levels or sets of gray levels into various colors. Prior to the development of the desktop computer, a similar technique was performed on microscopic images.

Color translation is the general process in which the differential properties of a specimen are converted to a differentially colored image for observation, recording, or data processing. According to Zworykin, Brumberg working in Moscow was the first to suggest the technique of color translation.

Several techniques produce color translation, including polarized light microscopy and DIC and other types of interference microscopy. In 1952 Zworykin used a vidicon television camera to make a color-translating television microscope. In 1960 he developed a special ultraviolet-sensitive image orthicon camera with a rapid readout and rapid erasure of the stored information during readout by the scanning beam. With this instrument, he examined a wide variety of biological specimens.

In 1949, Ridley developed a television ophthalmoscope to observe the retina. Ridley noted the advantages of television microscopy: multiple CRTs could be used for teaching purposes, with rotating trichromatic screens that allowed the image to be seen in true color, and the image contrast could be enhanced electronically.

In the middle of the twentieth century, optical microscopy was a standard research instrument in the laboratory. The problems of resolution and optical aberrations were solved and commercial microscopes of high quality became available from a variety of manufacturers. The development of epi-fluorescence microscopes and the concomitant advances in fluorescent dyes and fluorescent antibodies chemically attached to proteins resulted in very high specificity and sensitivity in optical microscopy. The net major advances occurred in the development of techniques to enhance the contrast of specimens devoid of inherent contrast; for example, living cells in culture. For the first time, phase contrast and differential interference contrast microscopy permitted cells in culture to be observed over time with light microscopy. These technical developments resulted in numerous studies of the dynamics of living cells: division, differentiation, fertilization, cell death, and disease development.

While differential interference contrast microscopy showed cellular images as "pseudo-three-dimensional objects," image interpretation in terms of cellular structures is both difficult and open to false interpretation. When microscopists attempted to observe thick, highly scattering specimens at high resolution, the result was very unsatisfactory. Similarly, the use of epi-fluorescence microscopy with thick fluorescent specimens gave blurred, low-contrast images within the specimen. The most modern and advanced techniques of optical microscopy could not be used with live, thick tissues and organisms, mainly because of the lack of depth discrimination of the conventional epi-fluorescence microscope. Part II of this book analyzes the various solutions to this formidable problem.

5.3 Summary

- In addition to providing adequate resolution, a light microscope must also provide sufficient contrast; otherwise the image would not be visible.

- The technique of **lifetime imaging microscopy**, in which the lifetime of the fluorescence forms the image, not the intensity as in standard fluorescence microscopy, is finding new applications in cell biology and medicine.
- A phase contrast microscope forms the phase contrast image by optically processing the direct and scattered light differently, then forming an interference pattern from both these waves. A phase contrast microscope converts differences in phase into differences in intensity in the image that can be detected by the eye or an electronic detector.
- In phase contrast microscopy, the resolution of the microscope is impaired since the full aperture of the microscope objective is not used.
- The principle of DIC microscopy is that the phase shifts resulting from the refractive structure of the specimen are encoded in a field of polarized light. The two superposed components are both offset and analyzed to show refractive index gradients that are converted into intensity differences in the image.
- The DIC microscope images the rate of change of the optical path difference. On the other hand, the phase contrast microscope images the absolute level of the optical path difference.
- As with phase contrast microscopy, the interpretation of the images obtained with DIC microscopy can be difficult and misleading, so caution is always advised.

Part II

Confocal Microscopy

Chapter 6

Early Antecedents of Confocal Microscopy

6.1 The Problem with Thick Specimens in Light Microscopy

It was evident to users of the light microscope that there were still unsolved problems with thick, highly scattering specimens. The use of the fluorescent light microscope together with fluorescent, thick specimens was difficult; moreover, light from above and below the focal plane contributed to a blurring of the image and a general loss of contrast. These problems were also evident during *in vivo* microscopy of embryos, tissues, and organs. On the other hand, these problems did not exist for very thin, fluorescent specimens.

Real biological specimens have internal structures that vary with depth and position. Prior to the use of three-dimensional computer reconstructions, in order to obtain a valid understanding of the heterogeneous specimen, it was necessary to use the light microscope to image many focal planes from the top to the lower surface, and then to reconstruct either a mental three-dimensional visualization of the specimen, or use computer techniques to make this visualization. This was the technique used by Ramón y Cajal in his seminal microscopic studies of the vertebrate nervous system.

The laser was invented by Theodore Maiman in 1960. Two years earlier, Arthur Schawlow, Charles Townes, and, independently, Alexander Prokhorov showed that it was possible to amplify stimulated emission in the optical and infrared regions of the spectrum.

The Minsky patent for his confocal microscope was issued in 1961. Prior to these milestones, there were many technical innovations that aimed to increase the resolution of the light microscope. The laser is not a requirement for the confocal microscope, since usable light sources include the sun, white light arc lamps, and a 12V halogen lamp. We now discuss some of these innovations and their role in the development of light microscopy.

6.2 Some Early Attempts to Solve These Problems

A series of creative technical innovations in the field of light microscopy resulted in technical improvements and a deepened theoretical understanding of confocal light microscopy. The basic advances will be briefly discussed and classified into common groupings: advances in fluorescence microscopy and in light sources and point scanning. It is interesting that there were both parallel developments and reinvention of technical advances in disparate fields, and these processes continued to occur throughout the development of confocal and multiphoton excitation microscopy. Inventions were patented, patents were contested, and the process continues into the present.

The prerequisites for the developments of confocal and multiphoton excitation microscopy were the technical advances of the fluorescence microscope. In particular, the work of Brumberg demonstrated how to use light from above or a vertical illuminator together with a beamsplitter (dichroic) and filters in the filter block to separate the excitation light path from the emission light path to the eyepiece or detector (Brumberg, 1959). *In vivo* fluorescence microscopy must use the technique of epi-fluorescence or vertical illumination, since the standard technique of transmission fluorescence microscopy is not appropriate for *in vivo* microscopy. Brumberg also pointed out that the fluorescence signal is very weak compared with the reflected light signal; therefore, it is critical to separate the two signals in order to be able to detect the very weak fluorescence.

Furthermore, Brumberg's *in vivo* microscope with vertical illumination had several types of microscope objectives. Some objectives were sharply pointed, designed to penetrate tissues and organs. Others were constructed with a flat surface that applanates the surface of the tissue or organ during *in vivo* microscopy. Since the tip of the applanating microscope objective was on the surface of the specimen, it was necessary to devise a technique to vary the focal plane within the specimen. Brumberg also solved this technical problem.

Brumberg developed a method to shift the focal plane of the microscope objective while the objective was stationary with respect to the organ or tissue under observation. He constructed a movable lens within the microscope tube that would be translated along the optical axis to change the position of the focal plane of the objective. He further stated that changing the tube length induces optical aberrations with a given objective; however, small displacements (up to 10 mm) would not exceed the permissible limits of aberrations. Brumberg's microscope has been rediscovered many times in recent years and incorporated into the microscopes designed by others.

Brumberg also discussed the problem of glare in the field of view during *in vivo* microscopy of thick specimens. He noted that the source of the glare was fluorescence of the layer of tissue above and below the focal plane. He reported that this glare reduced image contrast and makes the image less distinct. Brumberg's solution was to use an annular ring in the microscope objective to achieve dark-field illumination, together with a two-color light filter mounted in the pupil of the microscope objective. There was some loss of brightness, but this technique achieved a gain in image contrast.

The second set of technical advances came in the form of scanned image microscopy. First, two very different types of illumination must be explained: Bright-field (wide-field) illumination and point scanning. To explain these terms, more definitions are needed. **Bright-field microscopy** involves direct light passing through the objective aperture and illuminating the background against which the image is observed. In **dark-field microscopy,** direct light is prevented from passing through the objective aperture. The image is formed from light scattered by features in the object; the detail appears bright against a dark background.

Reflected-light microscopy uses illumination that falls on the object from the same side as that from which the object is observed. Alternatively, **transmitted-light microscopy** uses transmitted-light illumination in which the light passes through the specimen. In a **reflecting microscope,** the objective and condenser are catoptric (based on mirrors, not lenses). Such systems are free from chromatic aberrations and have fewer spherical aberrations than dioptric (lens-containing) refractive systems.

A **scanning optical microscope** is designed to scan the object plane or image plane in a raster pattern and illuminates the specimen point by point, in a process called **point scanning**. A **raster** is a pattern of lines or points applied to an area in either a regular or random manner. Light signals at discrete and uniform intervals of position from the object are detected. The image is thus built up serially. There are two methods of scanning: in **beam scanning**, the illuminating beam is scanned over a stationary object. Alternatively, the object itself is moved and the illuminating beam remains stationary in **object scanning**. These will be discussed in the section on confocal microscopy. A **flying-spot microscope** is an early form of scanning optical microscope in which the intense spot of light forming the raster of a small CRT was imaged onto the object plane of a microscope through an eyepiece and microscope objective. A photodetector following the condenser receives the light transmitted by the specimen and modulates the brightness of the display CRT that is synchronized with the scan.

In a flying-spot ultraviolet microscope, the image source is a high-intensity CRT that emits in the ultraviolet. A high-NA lens projects the light onto the specimen. The specimen is scanned with the light in a raster or other pattern by modulating the electron beam in the CRT. The detector can be a photomultiplier tube and need not be an imaging device.

To summarize, both bright-field and dark-field microscopes are types of wide-field light microscopes. A wide-field microscope illuminates the specimen in parallel, and an image of the specimen can be observed in the ocular. In contrast, a scanning optical microscope illuminates the specimen by point scanning. There is no image of the specimen in the eyepiece; the image is built up serially and displayed on a computer monitor.

6.3 Scanning Optical Microscopes: How Scanning the Illumination Reduces Light Scatter and Increases Contrast

Microscopists discovered several ways to mitigate the problems of glare and the resulting loss of detail and contrast in the image of thick, highly scattering specimens. They observed that the images appeared sharper and with increased contrast if monochromatic light was used for illumination, because there is no chromatic aberration. Monochromatic light at shorter wavelengths also increases the resolving power of the light microscope. That is the basis of the ultraviolet microscope, which uses quartz optical elements to transmit the ultraviolet light. Next I present the advantage of scanning the illumination. which reduces glare, scatter, and increases image contrast.

In 1952 Harold Ridley developed an electronic ophthalmoscope to examine the retina of patients in the clinic. A CRT with a lens was used to scan a bright beam of light into the retina. The reflected and scattered light from the retina was detected by a photoelectric tube, and the image was displayed on another CRT. Both the primary CRT that provided the scanned illumination and the image-forming tube were synchronized. On the image-forming tube, the image of the retina was serially built up. That is an example of a scanning optical microscope.

What motivated Ridley to develop a scanning spot optical microscope? In his experiments, he observed that spot scanning the retina resulted in much higher contrast than when he used bright-field illumination to illuminate the entire retina at once. While this example refers to the retina, similar observations were made by many microscopists on other specimens.

Consider this analogy and its role in explaining contrast: Imagine an auditorium containing 10 rows and 10 columns of seats, with one person seated in each chair. Assume, furthermore, that everything in the room is white—the chairs, the walls, the floor, and the ceiling; and all the people are dressed in white clothes and wearing white masks. Arrangements are made to take a group photograph inside the auditorium. After taking and examining the photograph, we notice that we are not able to discriminate whether people are present or the auditorium is empty. The reason is that there is no contrast in the photograph—everything is the same intensity of white. If we assume absolutely uniform illumination and identical reflectance of every part of the auditorium, then every region of the photograph will have identical intensity. The lack of contrast makes it impossible to determine the nature of the photograph.

We repeat the illumination and photography process with one important difference: each person removes the mask and dresses in different colored and patterned clothing. Now we take the photograph and are able to distinguish each person. The different features of the face, hair, body, and clothing provide differences in the intensity of the light reflected into the camera and images onto the film. This shows the critical importance of contrast.

Now let us use the analogy to explain wide-field microscopy. In this case, all the seats are occupied with people wearing very different clothing. In order to take the photograph, a special lighting system is used so that the light evenly illuminates every individual in every seat. Therefore, those individuals in the last row and in the first row receive the same amount of light. The shutter of the special camera is opened long enough to form the image of the entire group in the auditorium on the film. Note that every seat is illuminated at the same time, which is equivalent to a wide-field microscope; there is simultaneous illumination of every spot of the object and the image-forming light is detected over the entire field in parallel.

Next, several people around various parts of the room decide to smoke a cigarette. We quickly take a photograph—under the same conditions as above—before there is sufficient time for the smoke to diffuse throughout the auditorium. In this case, the smoke will scatter the illumination light and affect the illumination of every seat in the auditorium. That will degrade the image quality in the photograph and reduce the contrast in the photograph, which is equivalent to wide-field microscopy with local light scatter.

We repeat the analogy under conditions that simulate scanning optical microscopy or point illumination. In this case, we use a new type of illumination system that illuminates each seat serially; that is, first row 1 column 1, then row 1 column 2, and so on to row 1 column 10; then it jumps to row 2 column 1, then row 2 column 2, etc., until all the seats are sequentially illuminated. The shutter is opened just before the illumination system is positioned at row 1 column 1, and is closed just after the illumination system is positioned at the last seat in row 10 column 10. The photograph is built up serially. This is analogous to scanning optical or point-scanning microscopy. The disadvantage of scanning optical microscopy over the wide-field format is that the scanning takes more time.

We repeat the case above with a few people smoking. Assume the smoke is limited to the regions next to each smoker and does not diffuse throughout the auditorium. Now we repeat the sequential illumination. We find that the light scattered in the smoking regions of the auditorium only affects the image quality and the contrast in those regions. The images of individuals who are not located in the proximity of the smoke retain their image quality and contrast. That is the great advantage of scanning optical microscopy or point scanning. Scanning optical microscopy was developed prior to the invention of the confocal microscope and produced images with increased contrast compared to wide-field microscopes.

This analogy reinforces several characteristics of real-world optical microscopes. First, contrast is necessary to observe a specimen. Even when the microscope provides the appropriate resolving power, without contrast there is no image of the specimen. Second, in wide-field microscope the specimen is illuminated homogeneously with Köhler illumination; all illuminated spots of the specimen are imaged simultaneously. On the contrary, with a scanning optical microscope the illumination is serially applied to the specimen, spot by spot, and the image of the object is serially formed on an integrating device from each corresponding spot on the specimen. While each spot of the illumination on the specimen simultaneously forms a corresponding spot in the image, time is required to build up the entire image. Third, scanning optical microscopy results in improved image contrast in the presence of inhomogeneous scatter within the specimen.

6.4 Some Early Developments of Scanning Optical Microscopy

Perusal of the patent literature on the development of scanning optical microscopes demonstrates the clever implementation of mechanical devices. In the late 19th and beginning of the 20th century, the design and construction of precise mechanical devices reached a high level of excellence. Electrical, electromechanical, and finally electronic (first analog and then digital) devices and computers were inte-

Figure 6.1 Paul Nipkow.

grated into new types of microscopes. In the 1930s and 1940s, the invention and development of nonoptical microscopes such as the electron and field-ion microscope permitted the visualization of objects (e.g., viruses and atomic arrangements on crystal lattices) that were below the resolving power of light microscopes.

The confocal microscope was invented in 1957. The invention of the laser (1960) together with the development of the desktop computer resulted in the commercialization of confocal light microscopes and their widespread distribution and use. Prior to these two seminal inventions, many important developments occurred in scanning optical microscopy.

In 1884, Paul Nipkow (see Fig. 6.1) invented the electrical telescope, a forerunner of our modern television. The principal problem solved by Nipkow was how to construct a mechanical device to dissect an image of an object into many parts, transmit these parts serially over an electric wire, and, finally, reconstruct the original image. If these processes could be performed at a sufficient speed, Nipkow's electrical telescope (television or distant vision) could respond to real-time motion. The key component is the **Nipkow disk**, a rotating disk with holes arranged in a spiral or interleaved set of spirals. The rotating disk with its spiral sequence of holes is a simple mechanical method to scan a light beam over a circular lens region. The Nipkow disk was later to be used as the basis of beam- scanning real-time confocal microscopes.

The idea of scanning an image did not originate with Nipkow and his television. In 1843, Alexander Bain patented a system to scan images, dissect the image into parts, transmit the parts electrically, and then at a distant site reconstruct the parts into the image. That invention was the first fax machine.

In a 1951 paper by Robert C. Mellors, the utility of microabsorption spectra of naturally occurring biomolecules in thin smears of cells in both the infrared and ultraviolet regions was described. Mellors explained how a multimodal microscope could be used in the visible region with bright-field, dark-field, phase contrast, and fluorescence microscopy, and in the ultraviolet and infrared regions with a method to convert the resulting image into the visible range. The microscope could also perform microspectroscopy of various regions of cells and tissues. The only limitations were that the specimens had to be thin; consequently, they were usually smears of cells on a microscope slide. Thus, Mellors bypassed the difficult problem of fluorescence microscopy of thick specimens.

In 1951, Mellors and Reuben Silver at the Sloan-Kettering Institute for Cancer Research, New York, published "A microfluorometric scanner for the differential detection of cells: Application to exfoliative cytology" (Mellors and Silver, 1951). Their motivation was to develop a light microscope for the automatic searching and detection of cancer cells in exfoliated cells obtained from tissue secretion and spread on microscope slides. Their microscope used Köhler illumination; the cells in the smear were stained with a basic dye that selectively combines with the nucleus of each cell. The authors determined that the fluorescence intensity of cancer cells is two to three times greater than that for normal cells. In order to scan the fluorescence from each cell on the microscope slide, it is necessary to allow the fluorescence from each to fall separately and successively on the photocathode of the detector. That process is performed by the Nipkow disk. The rotating disk is placed in the intermediate image plane and forms a raster scan of the fluorescence light that is imaged with a lens onto the photocathode of a photomultiplier tube. The rotating disk contains a spiral of round apertures spaced at equal angular intervals. The size of the apertures is equivalent in the object plane to 10 μm, or the size of a cell nucleus. The amplified output of the detector is displayed on a CRT. The intensity of the fluorescence is used to discriminate normal from cancerous cells.

The microfluorometric scanner consists of a light source providing long-wave ultraviolet light, light microscope, Nipkow scanning disk, photomultiplier detector, and voltage pulse discriminator and counter. The most important scanner feature is the Nipkow disk; hence, the fluorescent light from each cell in the specimen is separately and successively detected, and therefore the intensity from each cell in the smear can be measured. The fluorescence microscope uses vertical illumination and the special dichroic filter deflects the ultraviolet light onto the back focal plane of the microscope objective, and separates the fluorescence from the excitation light, only passing the longer-wavelength fluorescence light to the Nipkow disk scanner and detector.

The Mellors and Silver scanning microscope may be considered to be half of a confocal microscope. The Nipkow disk scanner could be placed in the image plane to scan the image before the light was imaged onto the detector. Alternatively, their microfluorometric flying-spot scanner could be constructed with the Nipkow disk placed in the path of the illumination light. In that case, the fluorescence microscope would operate as a transmitted-light fluorescence microscope. The purpose of the Nipkow scanning disk would be to scan the illumination light serially over the specimen. What would be required to have a true confocal microscope? Both implementations would be combined and two Nipkow disks rotating synchronously would be required: one on the illumination side to illuminate the specimen with a small spot of light, and one on the image side, to limit the fluorescence light to a spot conjugate to the illuminating spot. Mellors and Silver came very close to inventing a true confocal microscope in 1951.

We now discuss an important instrument development that uses a CRT to provide a television raster scan of light over the specimen. This instrument contrasts with the previous scanning microscope developed by Mellors and Silver that used a mechanical Nipkow disk to provide scanning.

In 1952 F. Roberts and J. Z. Young (an anatomy professor at University College London) published their paper on a new flying-spot microscope (Roberts and Young, 1952). In it, they state that in the history of microscopy most of the information obtained has been qualitative. The idea behind a flying-spot microscope is the combination of a microscope with the well-known flying-spot video generator. That device was previously developed to scan a photographic plate and to convert the spatial distribution of density in the plate to a time-varying voltage that could be used for electrical transmission. A fax machine is a good example.

The authors then describe several important, unique applications of the flying-spot microscope with a CRT display. A video display could be used for educational purposes and the display could be distant from the microscope. They proposed applications in which the environment is dangerous; e.g., in atomic-energy research, where it is necessary to use the microscope with materials that are highly radioactive. Today, there are Internet-connected microscopes that permit both control of the microscope and viewing of the image in real time from all over the earth.

According to the authors, the biological utility of their flying-spot microscope depends on the high sensitivity of the photomultiplier detector and the ability to use electronic means for averaging and altering image intensity and contrast. The authors claim that these two features permit a shortened exposure to the damaging ultraviolet light in live-cell and tissue studies. They also describe the technique of color translation, suggesting that each cell and tissue constituent could be represented in a different shade of color. The flying-spot microscope has improved resolution, sensitivity, and contrast, and these enhancements also are applicable to microspectroscopy of cells and tissues.

The discussion of quantitative measurements was the forerunner of dedicated image-processing systems (using digital processing in a computer). Roberts and Young detail how to implement automatic counting, sizing, and sorting of microscopic particles. In addition, they suggest several clever applications for their instrument, including an electronic ophthalmoscope to image the retina (similar to the instrument used by Ridley), and a rapid data storage/retrieval system in calculating machines and computers (to replace the use of computer punchcards). They also state that the flying-spot microscope would be useful in fluorescent microscopy, to detect and count normal versus malignant cells and microorganisms versus pathogens.

In the flying-spot microscope of Roberts and Young, the light source is a flying-spot CRT that provides a small (25 μm on the surface of the CRT) spot moved at video frequencies. A microscope projects the moving spot onto the specimen. The detector is a photomultiplier tube. The microscope operates in transmission mode. Distortion in the image due to lens aberrations and afterglow in the scanner phosphor is electronically corrected. An amplifier increases the detector signal and corrects the system gamma. The display is a standard television or radar CRT.

The performance of the flying-spot microscope is excellent. Direct magnification of 500,000× can be obtained. The authors state that the four most popular methods of obtaining contrast are the use of selective dyes (staining), ultraviolet radiation, phase contrast microscopy, and polarized light. The authors claim excellent contrast with their flying-spot microscope even from the comparatively weak contrasts generated from unstained cells and tissues.

The depth of field, which is the thickness of the specimen that is in reasonably sharp focus, is stated to be the same as in the conventional microscope. However, the authors suggest that the use of optical sectioning to present a three-dimensional image would greatly increase the effective depth of focus.

In summary, the 1952 paper by Roberts and Young described raster scanning the specimen with a spot of light, detecting the transmitted light with a photomultiplier, displaying the image on a CRT or video monitor, electronically modifying the image-forming signal to increase contrast, using color translation to shade different specimen constituents various colors (the modern look-up tables for image display), and the use of optical sectioning.

Concurrent to their work, on the other side of the world, a confocal microscope was invented and constructed! Hiroto Naora at the Department of Physics, Faculty of Science, University of Tokyo, Japan, in 1951 invented a nonimaging confocal microscope for microspectrophotometry of the cell nucleus (see Fig. 6.2). His paper, "Microspectrophotometry and cytochemical analysis of nucleic acids," was published in the journal *Science* (Naora, 1951). The motivation was to improve the microspectrophotometry of cells, specifically to measure the spectral transmittance (an absorption versus wavelength curve) for a minute part of the cell nucleus.

The solution Naora invented was to use two identical microscope objectives: one placed below the specimen as part of the illumination system, and one placed above the specimen to collect the light from the specimen. Two apertures are placed in conjugate planes, one on the illumination side and one on the imaging side. The aperture on the illumination side causes the illumination of the specimen to be restricted to a very small focal volume. The light from this focal volume is passed through the second aperture on the imaging side and detected by the photomultiplier. However, all other light outside the focal volume will not be accepted by the imaging aperture and is thus blocked from detection. Therefore, stray light from above and below the focal plane is strongly rejected. The Naora confocal micro-

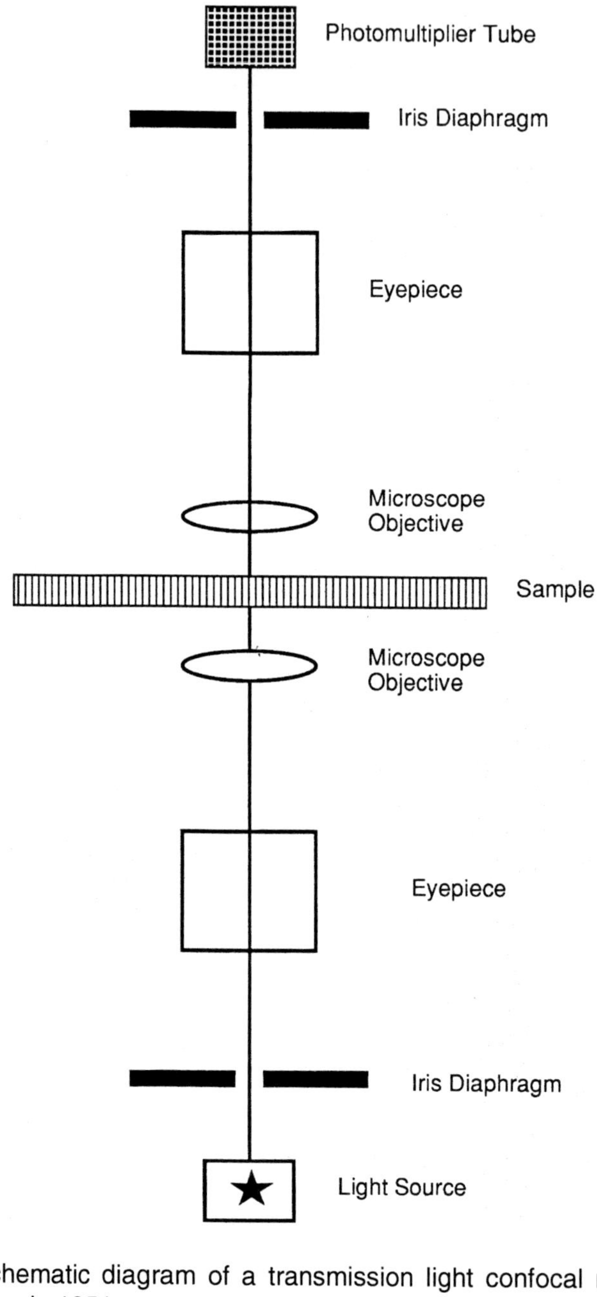

Figure 6.2 Schematic diagram of a transmission light confocal microscope invented by Naora in 1951.

scope has depth discrimination. This microscope, as designed and constructed by Naora, may be the first confocal microscope—albeit, a nonimaging instrument—to be described in a scientific publication.

In 1970, Klaus Weber, a scientist working in Wetzlar, Germany, at the Ernst Leitz Company, invented a "device for optically scanning the object in a microscope." The U.S. patent issued for this invention contains drawings and descriptions of several types of scanning confocal microscopes. All of these clever designs used oscillating mirrors for point scanning the specimen, and two conjugate apertures to discriminate against out-of-focus light. One embodiment of his invention uses two Nipkow disks that rotate on a common axis. Another embodiment uses the synchronous displacement of plates, containing either circular or slit apertures, in the illumination and detection paths.

When a plane mirror is placed on the stage of a standard light microscope and placed in the focal plane of the microscope objective, a sharp image is observed in the ocular. As the microscope is defocused by moving the mirror toward or away from the tip of the microscope objective, the image of the mirror becomes blurred. When the same procedure is repeated with a confocal microscope, an entirely different result is observed. As the confocal microscope is defocused, the image of the mirror disappears! A confocal microscope has depth discrimination, which is a result of its axial resolution.

A **confocal microscope** operates on the following principle. A light source placed behind a pinhole forms a point source of light that is focused on one spot of the specimen. The reflected light from that illuminated spot is imaged by a microscope objective and is focused on a pinhole in a plane conjugate to the illuminating pinhole. The second pinhole is placed in front of a detector. Both pinholes are located in conjugate planes and their images are cofocused at the specimen, which is the derivation of the word confocal. Only the reflected light that passes through the detector pinhole is detected and forms the image; the defocused light from above and below the focal plane of the objective is not focused on the pinhole and therefore is not detected. A complete image is formed by either scanning the illumination spot of light or the specimen in a raster pattern.

For purposes of consistency with the literature, especially the work of Wilson and Sheppard, this text uses the term conventional microscope to denote a nonconfocal microscope. The term conventional microscope includes both wide-field optical and scanning optical microscopes. If the context of the discussion is specific to either of these types of microscopes, then the type of microscope will be explicitly stated for clarity.

An example of three-dimensional confocal microscopy of *in vivo* human skin is shown in Fig. 6.3 (Masters et al., 1997). A thick, highly scattering living specimen can be optically sectioned into a stack of images. This stack of images, each with high resolution and contrast, can be reconstructed in a computer to form a three-dimensional digital image of the specimen. The three-dimensional image can be sliced with cutting planes of any angle and the resulting image displayed. When stacks of optical sections are acquired over a period of time, three-dimensional microscopy can be combined with a time axis to visualize time-dependent changes.

This chapter described some of the experimental approaches toward a confocal microscope. The concept of dissecting an image into discrete parts and recombining them to form an image was known for a long time. Similarly, scanning optical microscopes were perfected as their improved contrast became evident. Some of

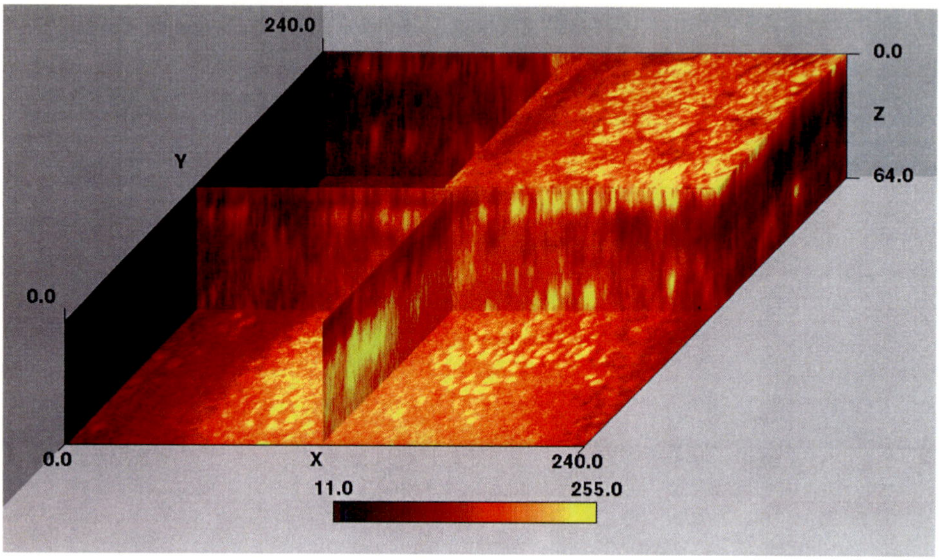

Figure 6.3 Three-dimensional reconstruction of in vivo human skin from images acquired with a real-time Nipkow disk confocal microscope; the reconstruction has a volume of 240 μm × 240 μm × 64 μm.

the inventions described in this chapter just missed becoming an imaging confocal microscope. Nevertheless, a continuous series of advances in fluorescence microscopy, scanning optical microscopy, and in particular the flying-spot microscope formed the foundation for the confocal microscope.

6.5 Summary

- Brumberg demonstrated how to use light from above or a vertical illuminator together with a beamsplitter (dichroic) and filters in the filter block to separate the excitation from the emission light path to the ocular or detector. This is the basis for epi-illumination fluorescence microscopy.
- Brumberg developed an *in vivo* microscope with an applanating microscope objective. Alternatively, a needle microscope objective was used to penetrate tissues and organs for *in vivo* microscopes. Within the microscope tube was located a movable internal lens used to shift the focal plane within the specimen.
- A scanning optical microscope was designed to scan the object plane or image plane in a raster pattern. Light signals at discrete, uniform intervals from the object were detected. The image is thus built up serially. There was a great improvement of image quality and contrast compared with bright-field microscopy. The limitation of a scanning optical microscope is increased image acquisition time.
- Images appear to be sharper and have increased contrast if monochromatic light is used for illumination.

- In 1952 the following aspects of modern microscopy were described in the paper by Roberts and Young: raster scanning the specimen with a spot of light, detecting the transmitted light with a photomultiplier, displaying the image on a CRT or video monitor, electronically modifying the signal that forms the image to increase contrast, using the technique of color translation to shade different specimen constituents various colors (the modern look-up tables for image display), and the use of optical sectioning to form a three-dimensional image with a great increase in the effective depth of focus.
- Hiroto Naora in 1951 invented a confocal microscope with two conjugate apertures, one on the illumination side and one on the detection side.
- The key difference between a conventional microscope and a confocal microscope is the optical sectioning (depth discrimination) capability of confocal microscopes. A conventional light microscope has no depth discrimination with defocus.

Chapter 7

Optical Sectioning (Depth Discrimination) with Different Scanning Techniques: The Beginnings of Confocal Microscopy

This chapter describes a variety of confocal microscopes, based on several types of scanning techniques, that were invented by different individuals in several countries around the world. Marvin Minsky invented a confocal microscope in which the specimen was mechanically scanned with respect to the illumination light. Petràn invented a confocal microscope based on the rotating Nipkow disk for scanning and descanning the light with respect to the specimen. Guoqing Xiao and Gordon Kino used a similar rotating Nipkow-type disk, but only used one side of the disk for scanning and descanning the light on and from the specimen. Svishchev used a two-sided mirror for the same process. Finally, laser scanning confocal microscopes are described.

In many cases, the motivation was a research problem that was not accessible with existing types of optical microscopes. Many of these problems were in the domain of *in vivo* microscopy. For example, Petràn investigated the living brain cortex and the live retina, and Svishchev studied the brain cortex in a living animal. These diverse inventors had a common requirement: an optical microscope that could image live, unstained, thick, highly scattering specimens. There is one striking exception. Minsky was attempting to use the light microscope to observe thick, fixed, Golgi-stained brain slices.

Instead of the low-contrast, blurred images from these specimens, these researchers dreamed of a microscope with depth discrimination that could be used to observe such specimens as the live brain cortex and living retina. Many of the early inventions of various types of confocal microscopes were driven by the limitations of existing optical microscopes.

7.1 The Confocal Microscope: The Problem and Its Solution

The lack of depth discrimination or optical sectioning capability is the major limitation of the conventional (nonconfocal) fluorescence microscope. In the past, the common solution was to use very thin specimens such as cells in tissue culture monolayers or thin smears of cells for pathology. Nevertheless, this limitation precluded the use of the light microscope for thick, highly scattering specimens, e.g., *in vivo* human skin, live embryos, intravital microscopy of organs, brain imaging, and studies of hard tissues such as teeth and bone. Similar problems occurred when such specimens were observed with reflected light in an optical microscope.

Wide-field fluorescence (nonconfocal) microscopy, when used with very thin sections under appropriate conditions, is still a very useful technique. When applied to thick, highly scattering specimens such as skin, the eye, brain slices or whole embryos, the inherent limitations become apparent. The lack of axial (depth) discrimination greatly degrades image quality. This is because of scattered and fluorescent light in the regions above and below the focal plane that are imaged together with the light from the focal plane.

Confocal microscopy offers a solution to this problem. The rejection of light from out-of-focus planes and the smaller depth of field result in images with high fidelity and high contrast. The confocal microscope also has the important ability to acquire optical sections from thick specimens. The solutions offered by a confocal microscope include enhanced axial and transverse resolution, enhanced contrast, and depth discrimination.

This chapter discusses inventions of various confocal microscopes. These differ from standard light microscopes in a critical manner. In the standard light microscope, the image blurs with defocus along the optical axis. In the confocal light microscope, the image becomes black with defocus; there is depth discrimination.

Another computational solution (not confocal microscopy) to the problem of depth discrimination exists: **deconvolution techniques**. Wide-field fluorescence microscopy can be used to acquire a stack of blurred images through the full thickness of the specimen. A measure of the actual axial resolution of the wide-field microscope can be made by imaging subresolution fluorescent particles such as submicron fluorescent beads under the same conditions as were used to acquire the stack of blurred images through the thick specimen. Various computer algorithms can deconvolve the blurred images and restore the image. These deconvolution techniques are not only of use with wide-field fluorescence microscopy, but may also help improve images taken with other types of light microscopy, e.g., confocal or multiphoton excitation microscopy.

The confocal microscope provides *en face* images of the specimen; hence, the plane of the image is orthogonal to the specimen thickness. For example, the confocal microscope, when applied to the skin surface of the arm, acquires images parallel to the skin surface; i.e., first the surface layer of cells, then the deeper cell layers. This is very different from the typical sections obtained in histopathology in which the tissue is cut along the thickness. In histopathology, a section of skin is removed, fixed and stained. For microscopic observation, the excised specimen is imaged in a plane perpendicular to the skin surface. Therefore, the microscopic image shows cells from the skin surface to the deeper cellular layers in a single image.

In contrast to the conventional light microscope, which images all of the points in the specimen in parallel, a confocal optical microscope optimizes illumination and detection for only a single spot on the specimen. In order to form a two-dimensional image with a confocal microscope, it is necessary to scan the illumination spot over the area of the specimen or to scan the specimen.

The next section describes and compares several generic types of confocal microscopes to explain their basic principles. That discussion is followed by the presentation of three designs that were based on modifications of the earlier instruments. Again, the reader is urged to read the original papers and patents.

7.2 Stage-Scanning Confocal Microscope Invented by Marvin Minsky

Marvin Minsky is credited with the invention and experimental realization of a stage-scanning confocal microscope. Minsky was motivated by his need to study the structure of fixed, Golgi-stained, thick brain slices. Minsky clearly stated the advantages of stage or specimen scanning in his 1961 patent on the confocal microscope. This idea decoupled the magnifications of the objective from the resolution. The magnification could be electronically varied by changing the number of pixels that form the scanned image. That implied that a single microscope objective with a fixed magnification could be used to form images of various magnifications. His patent also clearly showed the folded mode of modern confocal microscopes. Minsky described, but did not construct, a confocal microscope based on an **epitaxial** design, where the same microscope objective is used for both illumination and detection.

It is both instructive and of historical interest to follow the thinking of Minsky on his invention. First, he correctly stated the problem: How to make a microscope in which scattered light from a given point in the specimen is uniquely defined by a given illuminated point on the specimen. He realized that each focal point on the specimen would also have contributions from other points in wide-field microscopy.

He also realized that a second microscope objective could be used to illuminate one point of the specimen. That second objective replaced the usual condenser; it imaged a point source of light (obtained by using a pinhole aperture in front of a lamp filament). Now the illumination objective focused all the light from the point source (the pinhole aperture) onto a single point on the specimen.

Third, he noted that even with the second microscope objective illuminating a single point of the specimen with the image of a point source of light, the problem of scattered light from above and below the focal plane still existed. However, he noted that these out-of-focal-plane light rays could be eliminated by placing a second pinhole aperture in the image plane beyond the exit side of the microscope objective lens. This arrangement describes the principle of a confocal microscope.

The Minsky solution had elegant symmetry (see Fig. 7.1). There are two microscope objectives, one on each side of the specimen, and two pinhole apertures, one on the illumination side and one on the image side; therefore, both pinholes are located in conjugate planes. A point source of light illuminates a point on the specimen. The light scattered from that point is detected; hence, stray light from out-of-focus planes located above and below the focal plane is excluded (see Figs. 7.2 and 7.3). The word confocal denotes that fact: the images of these pinholes are cofocused or "focused together."

Minsky noted three other points. First, how can one build up an image from a series of single spots? The previous work on flying-spot microscopes solved that problem: the specimen could be moved in a raster scan pattern through the optical

Dec. 19, 1961 M. MINSKY **3,013,467**

MICROSCOPY APPARATUS

Filed Nov. 7, 1957

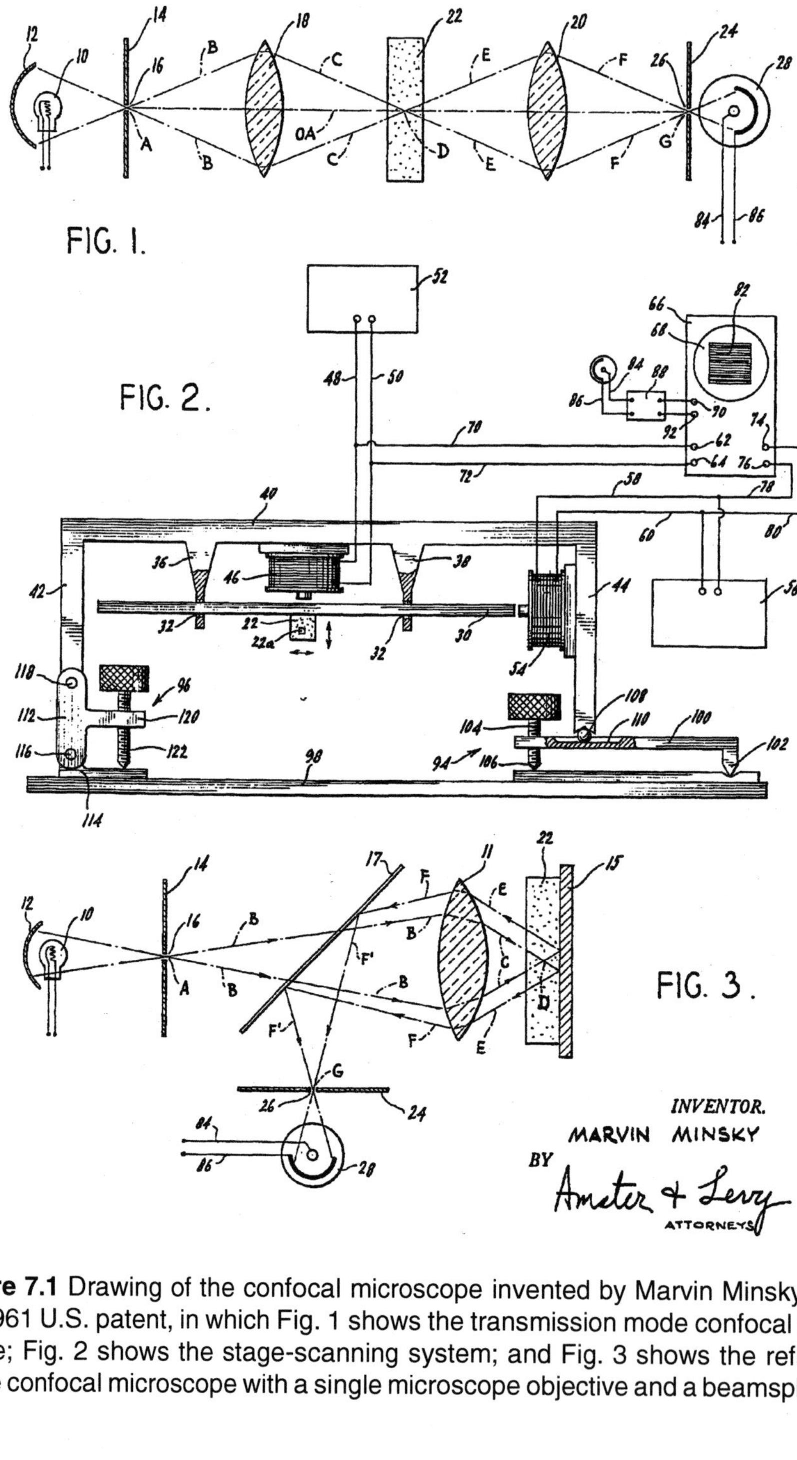

Figure 7.1 Drawing of the confocal microscope invented by Marvin Minsky, from his 1961 U.S. patent, in which Fig. 1 shows the transmission mode confocal microscope; Fig. 2 shows the stage-scanning system; and Fig. 3 shows the reflection mode confocal microscope with a single microscope objective and a beamsplitter.

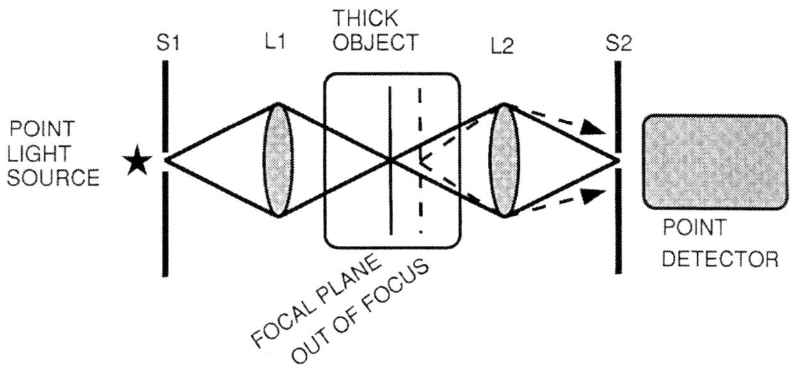

Figure 7.2 Schematic diagram of a transmission microscope illustrating the principles of a confocal microscope and depth discrimination, where S1 and S2 are confocal apertures located in conjugate planes, and L1 and L2 are focusing lenses for illumination and detection, respectively. The drawing is modified from Fig. 1 in the Minsky patent shown in Fig. 7.1.

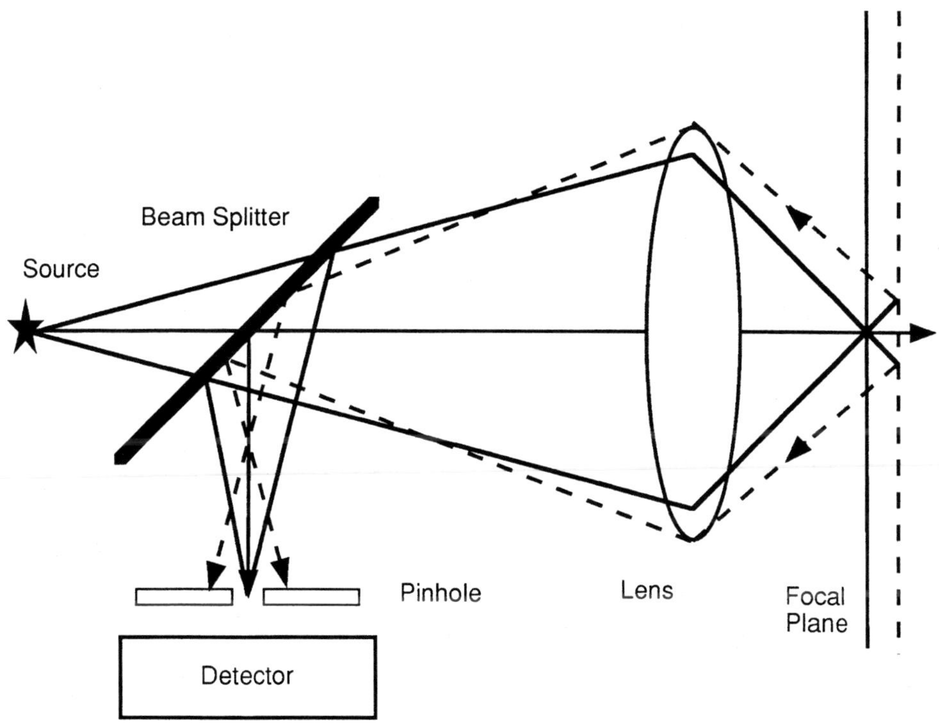

Figure 7.3 Schematic diagram showing the depth discrimination capability of a confocal microscope and how it discriminates against reflected light. The dashed vertical line to the right of the focal plane represents an out-of-focus plane.

axis of a stationary microscope, and the image would be built up, spot by spot, to make the complete image. The detector could be an integrating device such as photographic film or an electronic CRT display.

Second, the first design that we discussed operates in the transmission mode; he also described a reflected-mode confocal microscope, which is similar to the modern confocal microscope. A single microscope objective was used both to illuminate a single spot on the specimen and to image the collected light from that point. It was used together with a pinhole aperture on only one side of the specimen. A half-silvered mirror was used to separate the illumination rays from the imaging rays. Minsky stated that the diffraction patterns of both pinhole apertures are multiplied coherently with an increase of both axial and transverse resolution.

Third, in order to obtain a three-dimensional image, which was his original goal, an image is obtained on the CRT display, and then the specimen is translated a small distance (microns) along the optical axis of the microscope and the next image is built up. This process is repeated until a stack of images is formed, and the stack or optical section could then be formed into a three-dimensional image using techniques of stacked sheets of plastic containing a single optical section. These techniques were known in the fields of x-ray crystallography and light microscopy where they were used to reconstruct thick tissues and embryos.

The 1955 invention of the Minsky confocal microscope was a breakthrough, but one major limitation was the slow image acquisition time, since the image was slowly built up spot by spot, and the stack of images required moving the specimen incrementally along the optical axis. Minsky used an arc lamp for the light source. The detector was a low-noise photomultiplier tube. The display was a long-persistence radar scope. The acquisition time for one image was 10 seconds.

Two major technological advances were not available at the time of Minsky's invention: the laser was not yet invented, and the desktop computer was not yet available. These two inventions had a great impact on the popularity of modern confocal microscopy. Not until 1983, when Cox and Sheppard published their seminal paper, "Scanning optical microscope incorporating a digital framestore and microcomputer," did the microcomputer became part of the confocal microscope (Cox and Sheppard, 1983).

In 1971, P. Davidovits and M. D. Egger published, "Scanning laser microscope for biological investigations," which combined a 5-mW He-Ne continuous wave laser with a confocal microscope (Davidovits and Egger, 1971). Another unique feature of their confocal microscope was that the objective scanned over the specimen to form the image. Their paper also pointed out the problem of using coherent light in wide-field microscopy, in which interference effects severely degrade the image. They stated that another advantage of point scanning is that coherent interference does not occur.

Nevertheless, Minsky spelled out all of the key principles of confocal microscopy and constructed a working confocal microscope! Many of his ideas are implemented in the basic designs of modern confocal microscopes. Therefore, it is valuable to discuss his ideas further.

Minsky noted that two types of scanning were available: scanning the specimen or scanning the beam of light. Minsky correctly noted that beam scanning is fast; nevertheless, it is easier to keep the optics fixed, and to scan the specimen. He also noted that the use of beamsplitters always results in a reduction in image brightness. Minsky's confocal microscope used 45× microscope objectives in air. It could resolve points closer than 1 μm.

Minsky pointed out an important advantage of specimen or stage scanning. The microscope only used the central part (paraxial optics) of the microscope objectives; thus no off-axis or lateral optical aberrations exist that need correction. Chromatic and spherical aberrations still required correction. Minsky also stressed the advantage of combining stage or specimen scanning with paraxial optics. In addition, with stage scanning, if the microscope objective is used on-axis, then field curvature of the object is unimportant.

7.3 Mojmir Petràn, Milan Hadravsky, and Coworkers Invent the Tandem-Scanning Light Microscope

While the Minsky confocal microscope illustrated many technical developments that would appear decades later in modern commercial designs, it did not have an impact on the life science community at the time. A very different result followed the invention of the tandem-scanning confocal microscope; the life sciences community became deeply interested in this new invention.

The modern development of the real-time tandem-scanning confocal microscope is credited to the 1965 invention by Petràn and Hadravsky. In 1964, Petràn, who was a qualified medical doctor, visited Dr. Robert Galambos' laboratory at Yale University. They discussed the need for a microscope that could study live, unfixed, unstained neurons in the brain. During this visit, the tandem-scanning reflected light microscope was conceptually developed. A year later at Charles University in Plzeň, Czechoslovakia, Petràn and Hadravsky constructed the first prototype of a tandem-scanning confocal microscope based on a Nipkow disk. **Tandem scanning**, or double scanning, is defined as simultaneous scanning in both illumination and detection (see Fig. 7.4).

Petràn and Hadravsky were interested in intravital optical microscopic imaging neurons in live brain tissue. At Charles University, Petràn used his tandem-scanning microscope to investigate the live retina and the live brain. He and his students combined optical *in vivo* imaging and electrophysiological techniques. It is of interest that Minsky had a similar motivation in the design of his confocal microscope; however, he designed his microscope to study the three-dimensional organization of fixed, Golgi-stained, thick brain slices.

In addition to intravital imaging of brain tissue, Petràn and Hadravsky were interested in microscopic imaging of the structure of other living tissues such as epithelia, capillaries, nerves, muscles and glands *in vivo*. This was the driving force for the development of their Nipkow disk confocal microscope. Petràn later brought his tandem-scanning confocal microscope to the U.S. and collaborated

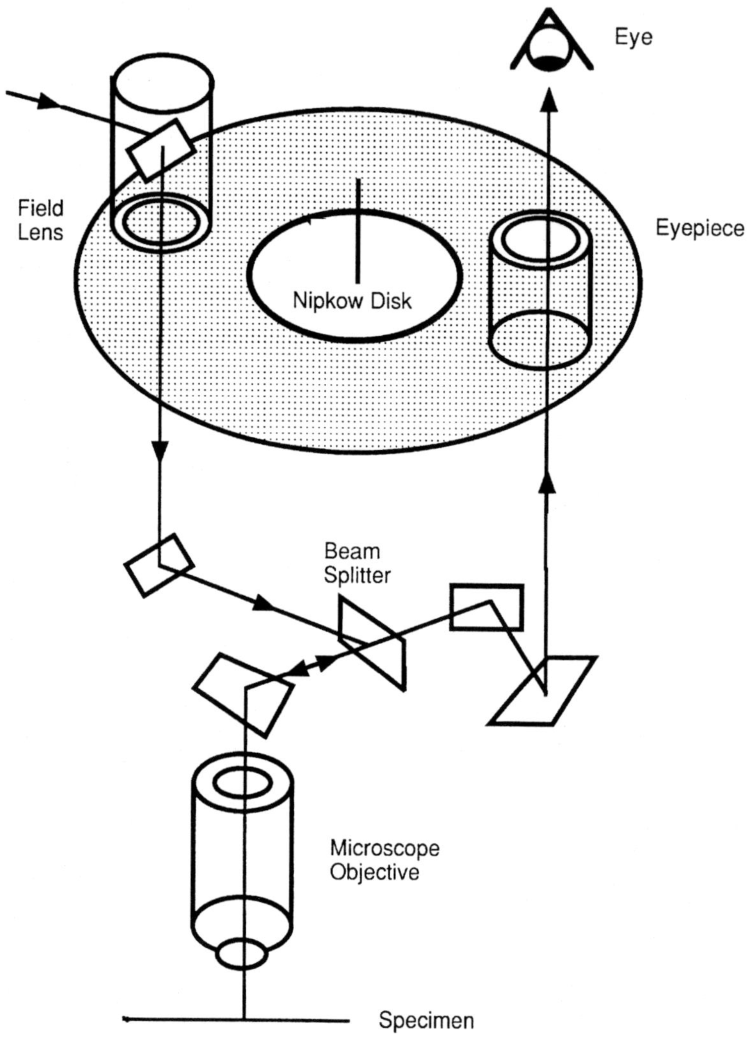

Figure 7.4 Schematic diagram of the real-time, direct-view, tandem-scanning Nipkow disk confocal microscope. The light source is a mercury arc or a tungsten filament lamp.

with Egger and Galambos at Yale University on experiments with live animals. Their 1967 paper was published in *Science* and included a composite hand drawing of the three-dimensional structure of a ganglion. In 1968 Petràn, Hadravsky, Egger, and Galambos published a paper on the tandem-scanning reflected-light microscope in the *Journal of the Optical Society of America*. That paper states that Galambos was the principal investigator of a NASA grant on the microscope, and that Petràn and Hadravsky were research associates at Yale University.

At that time, the small computers with three-dimensional volume rendering software that we have today did not exist. Therefore, there was not much interest in this technological development for the next 20 years. It is interesting to follow their formulation of the problem and then to appreciate their unique solution, which modified a very old invention.

A real-time tandem-scanning confocal microscope, in which the image could be observed with the naked eye, was developed by Petràn and Hadravsky in the mid-1960s. They acknowledged the contribution of Nipkow, who in 1884 invented the disk that provided real-time, tandem point illumination and detection. Petràn and Hadravsky invented their confocal microscope while living in Czechoslovakia, which had severe restrictions on travel, communication, instruments, and equipment. Therefore, great credit is due Alan Boyde, who aided them in the development and publication of their work. This shows the importance of free communication and free travel for the development of science and technology.

In order to understand their contribution, it is helpful to first formulate and state the problem: how to design a confocal light microscope that was simple, inexpensive to construct with materials then available, would use either the sun (it was first tested on a mountain) or an arc lamp as the source and a Nipkow disk for mechanical scanning, operate in real-time, and result in a real-color image.

Early on, the inventors decided against a transmission-light confocal microscope. A reflected-light microscope has several advantages. First, specimens such as whole animals, tissues, or organs could not be observed with a transmission microscope without sectioning; therefore, intravital microscopes would be excluded. Second, transmission confocal microscopes required two identical flat-field microscope objectives, which are difficult to obtain. Third, transmission light microscopy offers a high background of illumination. For these reasons, they decided to construct a tandem-scanning confocal microscope that operated in the reflection mode. Their patent, however, also describes a confocal microscope based on a spinning Nipkow disk constructed in the transmission mode.

The spinning Nipkow disk is the key component. The design concept was to have simultaneous point illumination of the specimen and detection of light from the same point. In order to form a two-dimensional image, a scanning device was required to simultaneously scan the image of both the illumination aperture and the cofocused image of the detection point over the specimen. A point-scanning confocal microscope suffers from the fact that the image is formed point by point. Petràn required a confocal microscope that would work in real-time. The use of a rotating Nipkow disk provided a mechanical device to permit the parallel illumination of many points on the disk; in effect, many confocal microscopes that work in parallel. At each pinhole on the illumination side of the disk, the light is focused by the objective to a diffraction-limited spot on the specimen. The light reflected from the sample is passed through a conjugate pinhole in the disk and can be observed in the eyepiece. When the Nipkow disk is rotated, a real-time image of the specimen can be observed.

The principle of the tandem-scanning confocal microscope is as follows (see Fig. 7.4). In the hypothetical case of only two pinholes on a stationary disk, the illumination light passes a pinhole on a stationary Nipkow disk and is focused by the microscope objective onto a spot on the specimen. The scattered and reflected light from that illuminated spot is collected by the aperture of the microscope objective. A very thin beamsplitter separates the illumination light from the reflected light. The reflected light is focused on a second pinhole, which is located in a position conjugate to the first pinhole; the images of both pinholes are cofocused on the specimen. The first illumination pinhole and the second imaging pinhole are located on a diameter of the disk at conjugate points. Only the light from the focal point in the specimen is focused on the second pinhole, and can therefore pass through the pinhole and form an image in the ocular. Light that is from above and below the focal plane of the objective is defocused and does not pass through the second pinhole. That is the origin of the axial discrimination in the tandem-scanning confocal microscope.

The idea of Petràn and Hadravsky was to pass the illumination light through one set of pinholes on one side of the Nipkow disk, and to pass the light from the specimen through a conjugate set of pinholes on the opposite side (see Fig. 7.4). This arrangement provided a solution to the problem of reflected light from the top surface of the Nipkow disk. The design of the tandem-scanning confocal microscope requires that the distribution of apertures have a center of symmetry, which results in identical aperture patterns in both the illumination and image fields.

In the actual microscope, the Nipkow disk contains many spiral arrangements of holes. Each aperture is in the range of 30 to 80 μm in order to avoid cross talk. About 100 pinholes at a time are illuminated on one side of the Nipkow disk, and the same number of conjugate holes pass the reflected light from the specimen. When the disk is stationary, the observer sees many spots of light from the specimen; when it rotates, the real-time image of the specimen is observed in the eyepiece. In addition to the microscope objective, beamsplitter, and Nipkow disk, a number of mirrors and lenses are contained in the microscope.

The first designs used several mirrors in which reflecting surfaces were perpendicular to the optical axis of the microscope. In order to reduce reflections from the microscope itself, it was necessary to use polarizers. Later designs of the direct-view tandem-scanning confocal microscope used prisms for beam inversion, and reduced the reflections from surfaces in the optical path. For example, the 1967 *Science* paper by Egger and Petràn describe a Nipkow disk confocal microscope with a polarizer, analyzer, and a quarter-wave plate to reduce reflections from the optical surfaces within the microscope. Prisms were used inside the microscope. For some of the experiments, the authors used the sun as a light source.

Petràn and Hadravsky decided to use multiple-aperture (multibeam) scanning since that would reduce the frame time to scan the field as compared to single-point scanning. The Nipkow disk contains several sets of pinholes (30–80 μm in diameter) arranged in several sets of Archimedes spirals. Each pinhole on one side of the disk has an equivalent and conjugate pinhole on the other side. The illumination light passes through a set of pinholes and is imaged by the microscope objective to form a diffraction-limited spot on the specimen. The reflected light from the specimen passes through a conjugate set of pinholes on the other side and can be observed in the eyepiece. Both the illumination and the reflected light are scanned in parallel over the specimen to generate the two-dimensional image of the focal plane by spinning the Nipkow disk. This microscope is called a **tandem-scanning reflected light microscope** since both (double or tandem) conjugate pinholes, located on opposite sides of the disk diameter, operate together or in tandem.

The user will notice that there is a large loss of signal in the confocal microscope designs that incorporate a Nipkow disk; since the ratio of the areas of the holes to the area of the disk is usually only about 1–2%, only a small fraction of the illumination reaches the specimen. This loss of signal is even more apparent when the Nipkow disk-based confocal microscope is used to image specimen fluorescence. Typically, the intensity of a fluorescent specimen is much lower than that from a highly reflecting specimen (i.e., semiconductors, hard tissue, mineral) and the fluorescent images are extremely weak. Therefore, the illumination must be very bright (a xenon or mercury arc lamp is usually required). Historically, to test the first instrument and in further work at Yale University, the sun with a heliostat on the roof of the laboratory was used as the light source.

These tandem-scanning confocal microscopes based on a Nipkow disk are best suited for reflected light confocal imaging. But, even in the reflected light mode, confocal microscopes based on a Nipkow disk containing pinholes have a very poor light throughput. In order to minimize cross talk between adjacent pinholes on the Nipkow disk, it is usually designed so that the separation between adjacent pinholes is about 10 times the pinhole diameter.

Various designs of the tandem-scanning Nipkow disk-based confocal microscope have been made. It is possible to make the Nipkow disk with several bands of apertures of varying sizes that are placed in the beam path. Real-time, direct-view scanning Nipkow disk confocal microscopes use round holes in the spinning disk; however, other designs have used square and rectangular holes. Another design used a spinning disk with slit apertures. In fact, in 1969 Egger, Gezari, Davidovits, Hadravsky, and Petràn designed and constructed a confocal microscope based on a rotating disk with slit apertures.

A tandem-scanning Nipkow disk-based confocal microscope is a poor choice for weakly reflecting specimens such as living cells, tissues, and organs. It is also not suitable for imaging weak autofluorescence or weakly stained fluorescent specimens. The low intensity of light that reaches the detector results in an image with marginal quality. However, for strongly reflecting objects such as hard tissue, composites, and microelectronics, the use of this type of confocal microscope is reasonable.

The advantages of the Nipkow disk-type confocal microscope are that it allows for real-time viewing, true specimen color, and direct observation. The microscope can also be used with white light and a microscope objective specifically selected because of its large chromatic aberrations. When the profile of a surface is to be imaged, the chromatic aberrations in the objective will separate the focal planes of light in the specimen according to the wavelength (color) of the light, and the resulting image will resemble a topographical map of the surface with different heights encoded into different colors. Finally, the clever design of tandem scanning presents a simple mechanical solution that incorporates point scanning to form the image, and the use of sets of conjugate apertures to strongly discriminate against light from above and below the focal plane provides axial discrimination.

In conclusion, the real-time direct-view scanning confocal microscope based on a spinning Nipkow disk is an elegant solution to the inadequacies of previous wide-field light microscopes. With bright or highly reflecting specimens, the images are seen in real color and with excellent contrast. The microscope design is simple and it can be manufactured at low cost.

The limitations of this type of microscope are evident when the specimen has weak fluorescence. The size of the disk apertures is fixed; nevertheless, several sets of apertures with varying sizes could be located on the Nipkow disk. Also, the mirrors of the tandem-scanning confocal microscope are difficult to align and maintain correctly. Mechanical vibration that causes the disk to wobble while rotating can degrade image quality and brightness. Also, the large number of optical surfaces, in which each contributes to the loss of light throughput to the detector, reduces the image brightness.

Petràn also suggested that the use of an image-intensified video camera sensitive in the infrared would yield an additional advantage as a detector. The use of infrared light as the illumination would permit increasing the penetration depth within the specimen because of the reduction of scattering at the longer wavelengths (compared to visible light) as the light penetrates the tissue, and also as the reflected light passes through the tissue toward the microscope objective. In Part III, we shall again see the utility of illuminating the specimen with infrared light and the concomitant increase in penetration depth.

The next two sections provide innovative solutions to the problems associated with the tandem-scanning confocal microscope: the one-sided disk and the Nipkow disk confocal microscope with a microlens array.

7.4 Guoqing Xiao and Gordon Kino Invent the One-Sided Confocal Scanning Light Microscope

The two-sided, or tandem-scanning, Nipkow disk confocal microscope is optimally suited for the observation of reflecting specimens; it solves the problem of stray light reflected from the top surface of the disk. However, it is difficult to adjust and align the microscope due to the number of internal mirrors or prisms and to wobble.

The idea of using only one side of the Nipkow disk for a confocal microscope was first suggested by Egger and Petràn. With this instrument, there are no mirrors below the Nipkow disk that must be aligned. In spite of this great advantage, they subsequently decided against it because of the serious problem of eliminating the light that was reflected from the top surface of the disk.

Other groups also explored the one-sided Nipkow disk confocal microscope. In 1975 Albert Frosch and Hans Erdmann Korth were granted a patent that they filed in 1974 for a "Method of increasing the depth of focus and or the resolution of light microscopes by illuminating and imaging through a diaphragm with pinhole apertures." Their invention was based on a **one-sided Nipkow disk** and they described a method to eliminate light from the disk, part of which involves tilting the Nipkow disk with respect to the optical axis of the microscope.

Xiao, Corle, and Kino, working at Stanford University, invented a real-time, one-sided, Nipkow disk-based confocal microscope (see Fig. 7.5), for which they

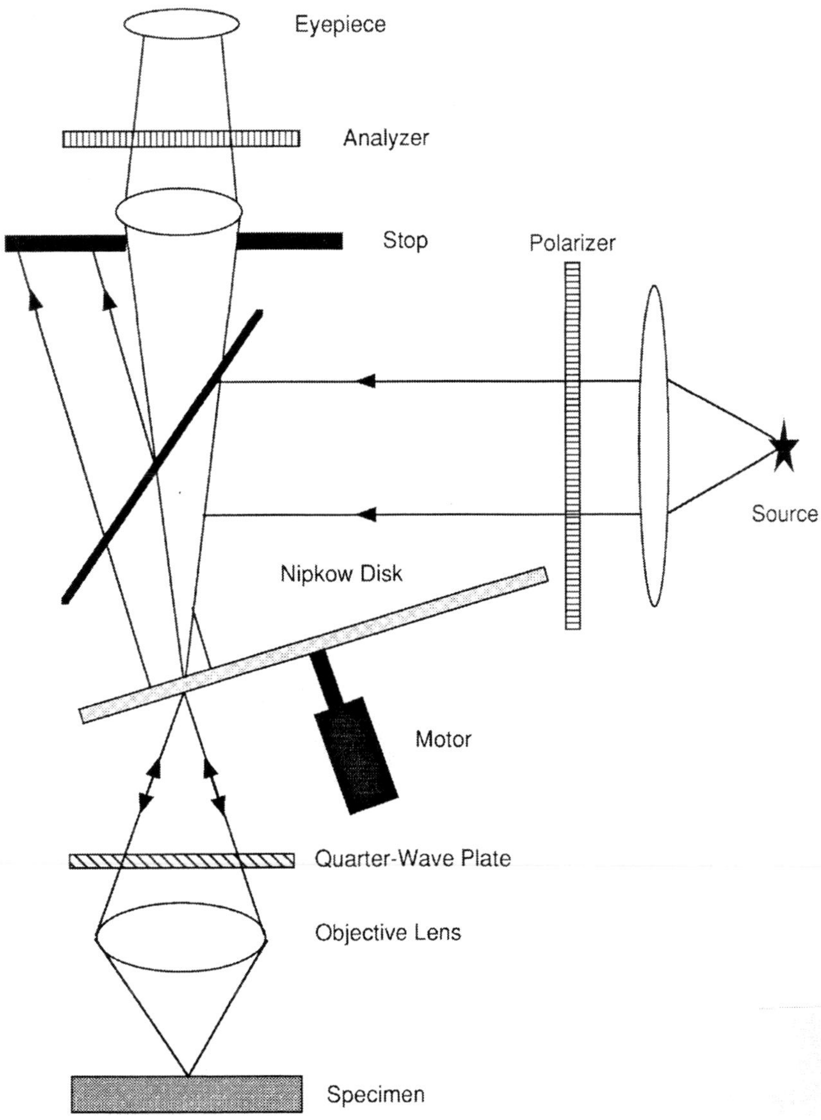

Figure 7.5 Schematic diagram of a real-time, single-sided, Nipkow disk confocal microscope. The actual Nipkow disk is about 10 cm in diameter and contains 200,000 pinholes, which are typically 20 μm in diameter.

received a patent in 1990. This design has several advantages over the tandem-scanning confocal microscope: it is less sensitive to vibration, has a simplified optical design, and is easier to align. Still another advantage is that while the disk is rotating, it can be translated horizontally, so different bands containing different pinhole sizes and/or shapes will be placed in the light path. This is a technique to change the pinhole or slit size without removing the disk and replacing it with another. This feature can be useful in the observation of specimens that have regions of differing reflectivities.

The driving force for their invention was the need to improve the metrology of semiconductor devices using simple optical confocal microscopes. Their confocal microscope used a rotating Nipkow disk in which the illumination and the reflected light passed through the same holes.

To reduce the reflected light from the surface of the Nipkow disk, three techniques were implemented. The disk was tilted approximately 5 deg. from the optical axis, so that the light reflected from its surface was reflected into a beam stop. The surface of the disk was blackened to reduce surface reflections. A polarizer was placed between the light source and the disk; hence, the disk was illuminated with polarized light. A quarter-wave plate was placed between the Nipkow disk and the microscope objective, and an analyzer between the Nipkow disk and the detector. The combination of polarizer, quarter-wave plate, and analyzer effectively separates the light from the specimen and that reflected from the disk surface. This optical arrangement sharply discriminates light reflected from the surface of the disk; similarly, it slightly reduces the light reflected from the object that reaches the detector.

A disadvantage of the one-sided design is that since the illumination and reflected light follow the same optical path, it is not easy to correct for chromatic aberrations in the microscope. This design, as with the tandem-scanning Nipkow disk-based microscope, still has the disadvantage of the low disk transmission, which also makes the microscope a poor choice for weakly reflecting specimens. Neither the one-sided nor the tandem-scanning Nipkow disk confocal microscope are suitable for use with weakly fluorescent specimens.

7.5 Effect of Pinhole Size and Spacing on the Performance of Nipkow Disk Confocal Microscopes

I have discussed the arrangements of the pinholes in the Nipkow disk and their various shapes. What about their size and spacing? How does the size and spacing of the pinholes affect their performance?

In previous discussions I introduced the words "point source of light" and "pinhole aperture." An example of a point source of light is a star as observed by a telescope. The operational definition of these terms is that the geometrical image of the aperture is much smaller than its Airy pattern. Another way to frame the concept is as follows: if the geometrical size of a pinhole is less than the point spread function of the lens, then it can be considered a point pinhole. Sometimes in the literature on confocal microscope we find the words "point" and "spot" are interchanged. A point is a geometrical object with no dimensions or extent.

In a Minsky-type of confocal microscope, the images of both the source and the detector aperture are co-focused on the specimen. The tandem-scanning and the one-sided confocal scanning microscope were invented in response to the long image acquisition time associated with the Minsky type of microscope. The Nipkow disk confocal microscopes reduced the image acquisition time and thereby permitted real-time imaging by using multiple apertures—and therefore multiple beams—in parallel.

The designer of a Nipkow disk confocal microscope has several choices in the construction of the aperture disk. First is the consideration of the optimal diameter of the disk apertures. Second, there is pinhole spacing or the distance between adjacent apertures. Third is the aperture shape. Fourth, the aperture-disk designer must select the pattern of the apertures.

All these design parameters depend on the nature of the object to be observed. Are the objects highly reflective semiconductor devices or are they weakly reflecting objects? Are they highly or weakly fluorescent? Microscope performance depends on the specimen and on the criteria most important to the observer: for integrated circuits, the user is often interested in measuring the profiles of stepped surfaces; for biological applications, the user is more often interested in distinguishing two neighboring point reflectors or fluorescent points.

When selecting the optimal pinhole size, the basic rule is that the aperture size should be adequate for the resolving power of the microscope, i.e., smaller than the central intensity spot of the Airy disk on the eyepiece side of the objective. For example, the aperture size is about 20–30 μm when used with a 100×/1.3 oil-immersion microscope objective with light at a wavelength of 550 nm. The aperture size should be selected for a given microscope objective; however, if it is selected for a high-NA objective, then the effect of using a low-NA objective is not critical.

For a given wavelength of illumination and an aberration-free lens with a defined NA, a point of light will be imaged as the Airy diffraction pattern. The geometrical image of the source aperture should be smaller than the main intensity peak of the Airy pattern. When the apertures are too small, there is a loss of signal because of the loss of light intensity at the object. As the aperture size decreases, the ratio of the aperture area to the disk area decreases, and therefore illumination to the object is decreased. If the size of the apertures is too large, a loss of resolution occurs in both axial and transverse directions; also, the illumination may not fill the pupil of the microscope objective, which results in a loss of NA and a consequent loss of resolution.

Pinhole spacing is the second important design parameter. In the ideal Nipkow disk confocal microscope are many sets of conjugate points, one on the illumination side and one on the detection side of the disk. The images of these conjugate apertures are co-focused on the object. Cross talk, which occurs when light passes through apertures other than the conjugate apertures on the image side of the disk, is undesirable. This stray light from out-of-focus planes within the object, as well as from reflections in various parts of the microscope, will severely degrade the image contrast. When the Nipkow disk confocal microscope is used in the fluorescence mode, filters can separate the illumination and the fluorescence light based on their wavelength differences. When the pinhole spacing is too small, another effect occurs for the case of coherent light as the illumination source, e.g., laser sources: the appearance of speckle in the image resulting from the interference of light from adjacent points in the object.

7.6 Akira Ichihara and Coworkers at Yokogawa Institute Corporation Invent a Microlens Nipkow Disk Confocal Microscope

The fundamental limitation of the real-time, direct-view, Nipkow disk tandem-scanning confocal microscope is that the very small area of the disk that is covered with holes (typically 1–2%) results in an enormous light loss and low illumination efficiency. Therefore, it is difficult to use this microscope with weakly fluorescent specimens. This problem has been partially solved by a group of researchers of the Yokogawa Institute Corporation in Tokyo, Japan.

In the Yokogawa confocal microscope, a laser illuminates the upper spinning disk, which contains about 20,000 microlenses over the pinholes on the disk. The lower disk contains another 20,000 pinholes arranged in the same pattern as the microlenses on the upper disk. The key point is that the lower pinhole disk is localized in the focal plane of the microlens disk (see Fig. 7.6). The improved performance of this microlens confocal microscope is because of the enhanced "fractional area of the apertures," which results in increased illumination efficiency. Both disks rotate on a common axis. The light transmitted by each pinhole is focused by the microscope objective to a spot on the specimen. The reflected light from the specimen returns on the same path through the microscope objective and pinhole, and is reflected by a beamsplitter through a relay lens to a two-dimensional detector. The microscope uses the full NA of the objective. Approximately 1000 illumination beams of light are focused on the specimen at one time. This results in a brighter signal (because of improved illumination) and faster image acquisition.

Figure 7.6 shows the principle of the microlens confocal microscope. With the presence of the microlenses, the pinholes pass 40% of the light incident on the upper disk. The design achieves high light throughput and therefore high sensitivity even in the presence of weakly reflecting specimens. The small pinholes in the Nipkow disk achieve high resolution in the transverse and axial axes. Another advantage is the high frame rate: 1 frame/ms, though it is usually operated at video rates.

This clever microscope design has no optical relays between the pinhole and the objective lens. This is a great advantage for minimizing optical aberrations and distortions present in other designs. It also dramatically reduces the light loss so prevalent in other tandem-scanning Nipkow disk confocal microscopes. For a mi-

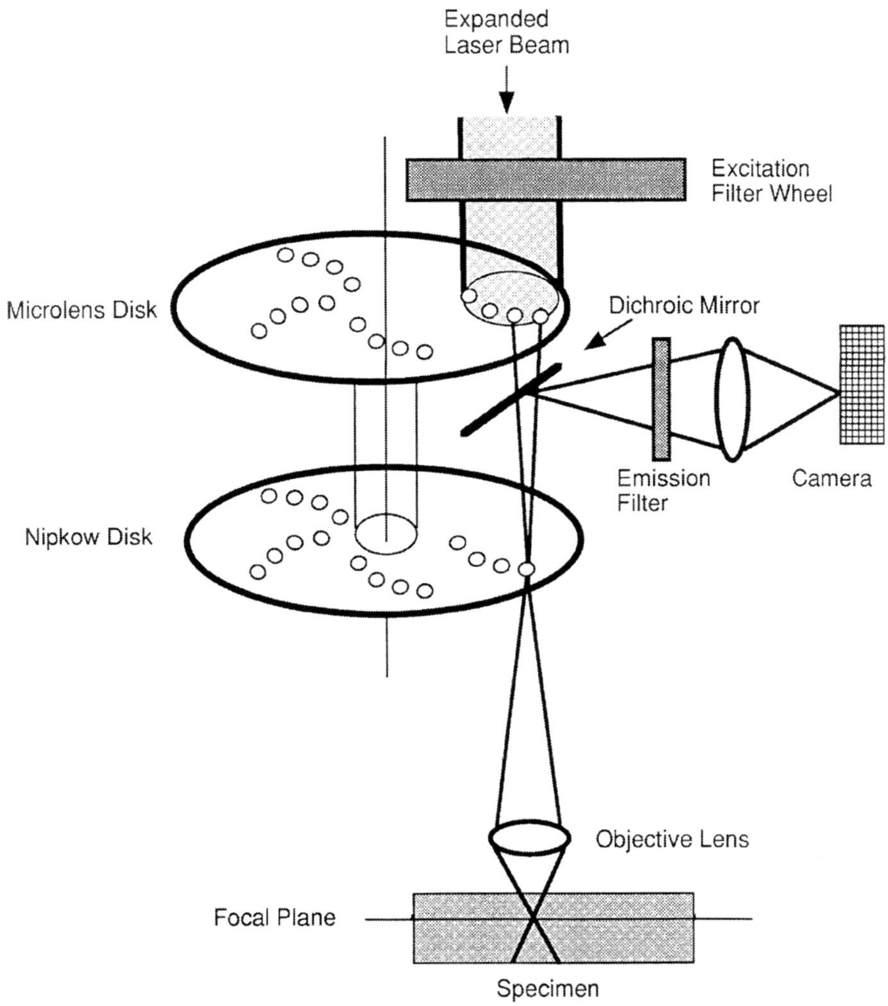

Figure 7.6 Schematic drawing of the microlens Nipkow disk confocal microscope.

croscope objective having an NA of 0.9 and a laser with a wavelength of 488 nm, the measured resolution on the optical axis is 0.6 μm (FWHM). Similar to the Kino design, a minimal effect of disk wobble on the image occurs. This type of confocal microscope is an alternative to laser-scanning confocal microscopes for studies on live cells and may have potential benefits for long-term cell biology studies. Now we leave the Nipkow disk confocal microscopes and discuss an alternative design based on conjugate slits.

7.7 Svishchev Invents an Oscillating Mirror Scanning-Slit Confocal Microscope

G. M. Svishchev, who worked at the Optical Laboratory, Institute of Biophysics of the Soviet Ministry of Public Health, Moscow, and was driven by the need to investigate neural tissue *in vivo*, invented a scanning-slit confocal microscope. Note that Minsky's confocal microscope was invented to study the three-dimensional structure of fixed, stained, thick brain slices. Petràn and his coworkers were also motivated to develop a new type of light microscope that could investigate the three-dimensional structure of unfixed, unstained, living brain slices. It is striking to note that different inventors, working independently in different parts of the world to achieve a common goal, invented three types of confocal light microscopes.

The key development in the Svishchev invention was a scanning system based on an oscillating two-sided mirror (see Fig. 7.7). This design is both simple and elegant; consequently, it eliminates the need for precision-controlled galvanometer mirrors for scanning and descanning. A simple two-sided mirror mounted on an oscillating rod performs the synchronized scanning, descanning, and rescanning for viewing. First published in a Russian journal in 1967 (and translated into English in 1969 and 1971), this clever design was subsequently redeveloped and reinvented in various designs of confocal microscopes in Europe and America.

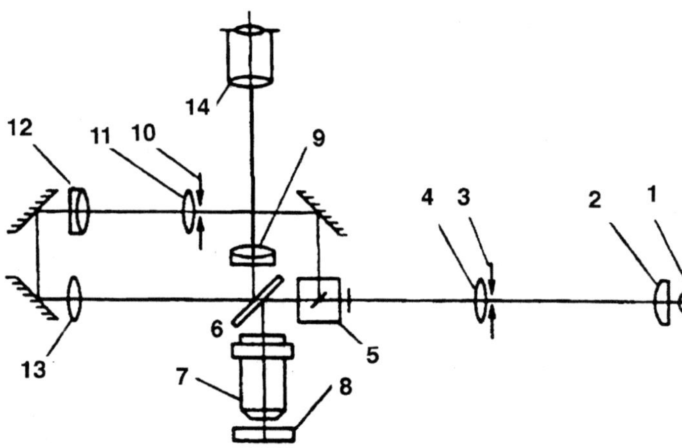

Figure 7.7 Schematic diagram of the Svishchev two-sided, oscillating mirror, scanning-slit confocal microscope. The light source, 1, is projected by a condenser lens, 2, onto the first slit, 3; the light passes through a prism cube, 5; an image of the first slit is scanned over the back focal plane of the microscope objective, 7, by the two-sided oscillating mirror, 6, which descans the reflected light from the focal plane in the specimen. The second slit, 10, is conjugate with the first slit (confocal) and excludes the light that is not in the specimen's focal plane, 8. The two-sided oscillating mirror, 6, performs three functions: it scans the image of the slit 3 over the back focal plane of the objective, descans the beam from the object, and rescans the beam for observation in the ocular, 14, or the film plane. Parts labeled 4, 9, 11, 12, and 13 are lenses.

Svishchev's confocal microscope provided for the effective removal of the light scattered by all the sample layers except the focal plane (Svishchev, 1969, 1971). This design produced a significant increase in the contrast of the image observed. Svishchev used an 85×, NA 1.0, water-immersion microscope objective. The use of a high-magnification, high-NA water-immersion microscope objective was subsequently incorporated into the instruments of Masters and Thaer. Svishchev published high-contrast photographs of the fine structure of cells and tissues in reflected light without the use of stains or ultraviolet light.

Furthermore, in a vision of future developments, Svishchev suggested the use of point apertures instead of slits together with continuous-wave lasers and solid-state detectors for improved imaging of thick, transparent, light-scattering objects. These proposals in his paper predated the development of laser-scanning confocal microscopes.

An alternative to point scanning, as exemplified in the designs of Nipkow disk confocal microscopes, is to use an illumination slit that is scanned over the back focal plane of the microscope objective. Since many points on the axis of the slit are scanned in parallel, the scanning time is markedly decreased; it can operate at video rate. Also, scanning-slit confocal microscopes have superior light throughput compared with point-scanning Nipkow disk systems. The disadvantages are that the microscope is truly confocal only in the axis perpendicular to the slit width, and it provides lower transverse and axial resolution than a pinhole-based confocal microscope. This comparison is for the same illumination and reflected light wavelength and the same microscope objective in each case. Even so, for confocal imaging of weakly reflecting living biological specimens, the trade-off between lower resolution and higher light throughput is acceptable.

Several arrangements have been developed to provide scanning of the illumination slit over the specimen and the synchronous descanning of the reflected light from the object. The simplest is the Svishchev design of a two-sided mirror mounted on a single oscillating shaft, which is the technique used in several modern designs of real-time confocal microscopes with bilateral scanning.

Scanning-slit confocal microscopes have several other advantages. The slit width can be adjusted, which allows the user to vary the thickness of the optical section as well as control the amount of light that reaches the sample and of reflected light that reaches the detector. This is important for samples that are very transparent, which can be imaged with a very narrow slit width; more opaque samples require a larger slit width.

As an example: the basal epithelial cells of a normal *in vivo* human cornea cannot be observed with a tandem-scanning confocal microscope. However, corneal basal epithelial cells can be observed *in vivo* when examined with a real-time slit-scanning confocal microscope. The reason is that although the tandem-scanning confocal microscope has higher axial and transverse resolution, the very low light throughput of the disk does not transmit enough reflected light from the specimen to form an image on the detector (in a single video frame) with sufficient signal to noise and, therefore, contrast to show an image of the cells. The reason is the low illumination efficiency of the Nipkow disk.

Now we have completed our discussion of Nipkow disk confocal microscopes, innovative modifications such as one-sided Nipkow disk confocal microscopes, Nipkow disk confocal microscopes with microlens arrays, and scanning-slit confocal microscopes. Today the most common type is a laser-scanning confocal microscope. Its origin was in the Minsky confocal microscope and patent, as well as many clever inventions in confocal microscopy since the time of his invention. In the next section we discuss several of these.

7.8 Laser-Scanning Confocal Microscope Designs

In this section, the design of the modern laser-scanning confocal microscope is presented. Two terms are often used interchangeably: laser-scanning confocal microscope (LSCM) and confocal scanning laser microscope (CSLM); this text will primarily use the former term. The previous discussion introduced many features and components, while this section provides some of the missing details and presents the LSCM as a complete instrument. Figure 7.8 shows the design of the LSCM.

Minsky's patent pointed out the key design principles for a confocal microscope. A careful reading of these patents is instructive to understand the various solutions to similar problems. Many of the ideas cited in these patents were eventually incorporated into commercial confocal microscopes.

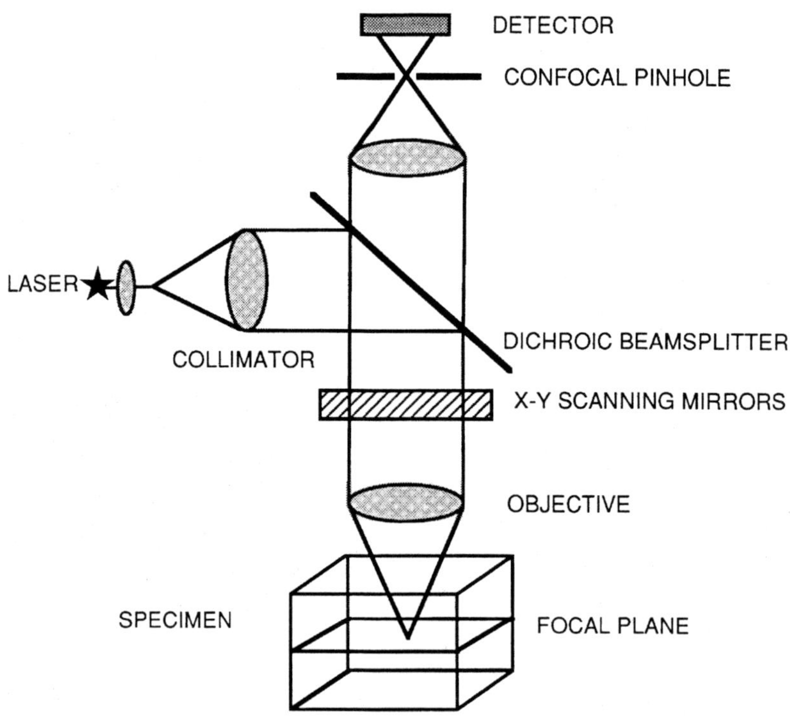

Figure 7.8 Schematic diagram of a confocal fluorescence microscope.

The design specifications of a commercial confocal microscope are constrained by many factors, but chief among them is intellectual property. Who owns the patents? Sometimes new designs were manufactured only to bypass the patents of others. Manufacturing costs and microscope maintenance is another consideration. The market is eventually heard, but change comes slowly.

Today, the user of a fluorescent confocal microscope has a choice of several excellent instruments. Many high-quality, corrected microscope objectives are available for a variety of purposes. Laser light sources are available to cover a range of wavelengths. The scanning systems and intermediate optics as well as the detectors are of high quality. These new confocal microscopes usually have several channels (wavelengths), a graphical user interface to operate and control the microscope, and software is available to process, analyze, and archive the images.

In the last few years, commercial laser-scanning microscopes have offered a variety of useful and improved features. New confocal microscopes are available with very compact scanning systems. They can be attached to a standard fluorescence microscope to convert it to a confocal microscope. A variety of air-cooled, low-noise compact lasers with a range of wavelengths can be purchased. Lifetime imaging can be added to these confocal microscopes by the use of add-on components. The graphical interface has been improved and a keyboard and mouse have replaced rotary dials.

While the original **analog-to-digital converters**, the electronic chips that convert the analog output of the photomultiplier detector to a digitized output, had a dynamic range of 8 bits, the new systems use 12, 14, or 16 bits. That results in a great improvement in the dynamic range of the system. A larger dynamic range permits more intensity steps between the lowest and highest light level in the image.

At first, the standard fluorescent microscope objectives were the only ones available. More recently, new series of microscope objectives have been specially designed for use with the fluorescent confocal microscope; the use of wider threads permits new microscope objectives with a high NA and a low magnification, which result in a wide field of view.

Confocal microscopes are also designed for clinical diagnostics. The goal of many of these is that the axial resolution will permit optical biopsy, that is, the diagnostic evaluation of tissue without the need for excision. There are confocal microscopes designed to image the skin *in vivo*; some use laser and others slit scanning.

In the field of ophthalmology, there are a variety of clinical instruments. Scanning-slit confocal microscopes that use noncoherent halogen lamp light sources are preferable for examining the cornea. Older designs are based on the Nipkow disk tandem-scanning confocal microscope. In addition, LSCMs are designed to image the cornea. Laser-scanning ophthalmoscopes are extremely useful diagnostic confocal microscopes for retinal examination.

The original patent of Minsky contained the concepts that are implemented in the commercial LSCMs used for both laboratory investigations and in the scanning laser ophthalmoscope. The availability of the laser provided a new, bright light source that resulted in several new laser-scanning microscopes. In the last decades, many technological innovations in beam-scanning confocal microscopes were developed. Wilson in Oxford, UK, and Sheppard in Sidney, Australia, developed various types of confocal microscopes. Brakenhoff demonstrated the importance of high-aperture immersion microscope objectives for optical sectioning.

The scanning-stage confocal microscope, which uses stage or specimen scanning, is another development that follows Minsky's original ideas. This type of confocal microscope uses the paraxial rays of the microscope objective, so the images are of excellent quality and contrast. The disadvantages are that the image acquisition time is slow (several seconds per frame) and the instrument is sensitive to vibration. For biological and clinical applications, the slow image acquisition time is undesirable; most users wish to view the image in real time as they vary the field of view and the position of the focal plane.

We now will review some interesting technological developments in components that preceded the modern LSCM. Full details are available in the original publications and patents. Not all of these technologies have become available in modern commercial instruments, although some of them have been implemented. Sometimes a company will approach the inventor and license the technology protected in a patent. Sometimes the company will market instruments based on the use of unlicensed intellectual property. In that case the patent owner may go to court; consequently, out-of-court settlements are not uncommon. In other cases, for example, with the patent covering the multiphoton excitation microscope, a company may go to court to challenge the validity of a particular patent. In the next paragraphs we briefly survey some of these key technical, patented developments.

In 1983, Werner Schmidt, Gerhard Müller, Klaus Weber and Volker Wilke, while working at Carl Zeiss-Stiftung in Oberkochen, Germany, invented a "Method and apparatus for light-induced scanning microscope display of specimen parameters and their distribution." One part of their invention contained all the components of a modern LSCM: a laser light source, point scanning of the diffraction-limited spot of light on the specimen, use of two orthogonal oscillating scanning mirrors that form the raster scan, and an aperture in front of the detector. Following the idea of Minsky, the Zeiss microscope used a folded design with a beamsplitter that operated in the reflected-light mode. What was new was the laser light source and the set of orthogonal scanning mirrors for laser beam scanning. The scanning microscope was designed to separate the illumination light from the fluorescence emitted by the specimen. It was also designed to simultaneously operate in two channels: e.g., Raman scattering and fluorescence, or scattered light and fluorescence. Their patent contained another important proposal: the light source could be an array of light sources (point sources in a linear or two-dimensional array), and the detector could be a linear array. The scanning and detection could be electronically synchronized by activating each single-point light source in the source array and the corresponding point detector in the detector array.

The next invention solved the problem of slow image acquisition that characterized the laser beam scanning confocal microscope based on two orthogonal, oscillating galvanometer mirrors (Draaijer, Houpt, 1988). Pieter M. Houpt and Arie

Draaijer in the Netherlands received a patent in 1989 for a "Confocal laser scanning microscope." Their goal was to design a laser scanning microscope with rapid line or frame scanning. They proposed to combine electronically a number of thin "optical sections" to form an image with an increased depth of focus. They proposed the use of an acousto-optical deflector together with a mirror galvanometer to increase the frame rate. (An **acousto-optical deflector** is a solid state device in which sound waves in a crystal form a diffraction grating that can be used to deflect the incident light. This phenomenon was predicted in 1921 by L. Brillouin for the case of a liquid traversed by ultrasonic waves and illuminated with white light. He predicted that diffraction would occur in analogy to a grating. In 1932 his prediction was experimentally confirmed by P. Debye and F. W. Sears, and independently by R. Lucas and P. Biquard.)

An acousto-optical deflector uses the first order of the diffracted light; hence, its wavelength and intensity are controlled by the frequency and amplitude of the ultrasonic wave in the crystal. The problem with acousto-optical deflectors is their wavelength dependence in both the deflection angle and efficiency. Therefore, longer fluorescent light cannot be passed back on the optical path through the acousto-optical deflector.

For the reflected-mode confocal microscope, the light is descanned from the specimen and the reflected light is passed through a pinhole aperture before detection. However, for work in fluorescence imaging, a dichroic mirror redirects the fluorescence light to a slit in front of the detector. Therefore, the fluorescence light does not re-enter the acousto-optical deflector. The fluorescence-mode microscope is only confocal in one dimension. The advantage is high image frame rates.

The use of galvanometer mirrors to speed image acquisition in confocal microscopes is embodied in the invention of Yoshiaki Horikawa, who worked at the Olympus Optical Company in Japan. His 1990 patent, "Scanning optical microscope," describes the invention: Light from a laser source is deflected by an acousto-optic deflector to scan the specimen at high speed; the light from the specimen does not pass back through the acousto-optical modulator, but is focused through an aperture to the detector. Since the diffraction in the acousto-optical light deflector is sensitive to wavelength, it cannot be used to descan the emitted fluorescence. If the device has a high diffraction efficiency for the laser light used for illumination, then the diffraction efficiency for the wavelength of the fluorescence is low. The result is a loss of fluorescence light. Another key part of the patent is that instead of a circular pinhole, a slit aperture is used in front of the detector. In summary, the light source is a laser beam that is scanned over the specimen. Scanning is provided by an acousto-optical device that scans the beam in the horizontal direction. An oscillating galvanometer mirror scans the laser beam in the vertical direction. The reflected light or the fluorescence light is taken out of the optical path by a beamsplitter, passes a slit aperture, and then is detected.

Another important advance was the 1987 publication by Kjell Carlsson and Nils Aslund, working at the Royal Institute of Technology in Sweden, of "Confocal imaging for 3-D digital microscopy," showing how the optical sectioning capability of the confocal microscope can be combined with digital image processing to provide three-dimensional microscopy. The inventors obtained a patent and constructed a beam-scanning confocal microscope that could be attached to a Zeiss Universal microscope, the only modification being a stepping motor on the fine focusing screw. The light source is an argon laser. Laser beam scanning is performed by placing two oscillating mirrors on orthogonal axes above the eyepiece of the light microscope. The slow-scan mirror is driven by a stepping motor, and the fast-linescan mirror by a galvanometric scanner. The laser beam is raster scanned over the back focal plane of the microscope objective. The light from the specimen is collected by the microscope objective, retraces the optical path, and is separated from the illumination light by a beamsplitter (dichroic mirror). The light from the specimen passes an aperture and is detected. Various dichroic mirrors are mounted on a rotatory wheel, and various pinhole apertures are mounted on a second rotatory wheel in front of the detector. The entire microscope system is controlled by a microprocessor. The confocal microscope could display 1024 × 1024 pixels, although it was sensitive to misalignment from vibration.

John White, William Bradshaw Amos, and Michael Fordham, working at the Medical Research Council Laboratory of Molecular Biology in Cambridge, UK, invented another variation of the LSCM. Their motivation was to produce a confocal microscope that was stable, insensitive to vibration, and optimized for fluorescence microscopy of biological specimens, mainly cells and tissues in culture. In addition, they wanted a design that could be retrofitted to an existing standard fluorescence light microscope. It was their 1987 paper, "An evaluation of confocal versus conventional imaging of biological structures by fluorescence light microscopy," that convinced the biological research community of the great utility of the LSCM.

Their invention is discussed in detail in their 1991 patent, "Confocal imaging system." In a 1991 correction to the U.S. patent, they list references to the missing U.S. Patent Documents—the patents of the following inventors: Barnes; Weber, Davidovits et al.; Baer; and Divens et al. Clearly, the White, Amos, Fordham design is based on many innovations of prior inventors. The unique feature of their invention is the nature of the input and detector apertures. An afocal set of mirrors transfers the beam from the first to the second galvanometer mirror. By expanding the optical path to almost 1.5 m, yet folding the optical path to make the instrument more compact, the apertures are millimeters in diameter. The optics formed an image about 80 times the magnification of the microscope objective at a distance of ~1.5 m. The Airy pattern at the detector was also enlarged. Therefore, a variable iris aperture from 0.7 to 7 mm could be used in front of the detector. Depending on the light intensity from the specimen, the iris could be made larger to pass more signal, or smaller to afford enhanced optical sectioning. An iris diaphragm has several important advantages compared to a pinhole aperture: it is adjustable, inexpensive, easy to align, and the dust and dirt that readily spoil a very small pinhole are not a problem.

A number of mirrors are used to fold the beam inside the laser scan assembly. Since each optical surface contributes to a loss of signal, the sum of all the surfaces from the folding mirrors, the two scanning mirrors, and the two concave mirrors of the afocal assembly result in considerable signal loss.

In the invention, the scanning elements are two plane mirrors oscillating on orthogonal axes. An afocal set of mirrors is another key component. These concave mirrors serve as a telescope to transfer the light from the plane mirror on the first galvanometer (y axis) to the plane mirror on the second galvanometer (x axis) and then to the eyepiece of the microscope. The scanning optics assembly is situated so that the exit pupil of the microscope eyepiece falls on the area of the plane mirror near its axis of oscillation. The use of spherical concave mirrors in the afocal assembly provides a system without chromatic aberrations, which is critical since the two- channel confocal microscope is used with at least two different fluorescent probes.

Other components of the scan head include a filter set adjacent to the laser source, and a second filter set that separates the light from the specimen into two channels, each with its own variable iris and its own photomultiplier tube.

Finally—and this is a key point in their patent—by adjusting the angle between the oscillating axis of the slow (frame) galvanometer mirror and the line of the input laser and returning beam, the scan lines on the specimen are linear.

7.9 Analytical Expression of Resolution in a Confocal Microscope

Lateral or **transverse resolution** is in the plane of the specimen or the x-y plane. **Axial resolution** is along the z-axis or the optical axis of the microscope. The lateral resolution of a confocal microscope is proportional to the NA of the microscope objective; however, axial resolution is more sensitive to the NA of the microscope objective. Therefore, to obtain the maximum axial resolution, and hence the best degree of optical sectioning, it is preferable to use microscope objectives with the largest NA. For an oil-immersion microscope objective with a NA of 1.4 and blue-light wavelength of 442 nm, the lateral resolution is 0.14 μm and the axial or depth resolution is 0.23 μm.

The theoretical analysis of the resolution in a confocal microscope expresses the resolution as a function of the wavelength of the light and the NA of the microscope objective. Another factor that affects the resolution is the contrast of the image. The contrast is determined by such factors as the number of photons detected per pixel, the noise contribution (Poisson statistics and instrument noise), the SNR, and the signal-to-background ratio. Resolution in a confocal microscope is linked to image contrast, and therefore to the number of detected photons.

We begin with the definition of a **resel**, which is the resolution unit transverse to the optical axis. The central bright portion of the diffraction pattern is the Airy disk, which has a defined radius of 1 resel. A real aberration-free microscope objective has a finite aperture, and therefore the resolution is diffraction-limited. If the lens has a NA that is given by $n \sin \Theta$, then a resel is defined as one-half of the diameter of the Airy disk, which is equivalent to the first dark fringe in the diffraction pattern.

The lateral resolution of a conventional and a confocal microscope are now compared following the analysis of Wilson (Wilson, 1990). We examine the case of a conventional microscope with the pinhole removed, and a confocal microscope with the pinhole placed in front of a detector. In each case, the image of a single point specimen is viewed in reflected light.

The following equation is also the analytical expression for the Airy disk. The Airy pattern is the response for an aberration-free lens with a circular pupil. This Airy pattern, also called the Fraunhofer diffraction pattern of a circular aperture, is what is observed when a point source, e.g., a star, is imaged with a telescope; it has a bright central disk surrounded by concentric bright and dark rings. The intensity of the bright rings decreases rapidly with their radius.

The image intensity of a conventional microscope is given by Eq. (7.1), which was derived by Airy in 1835,

$$I_{conventional}(\nu) = \left[\frac{2J_1(\nu)}{\nu}\right]^2. \tag{7.1}$$

where $I_{conventional}$ is the intensity of light from the object, J_1 is a Bessel function of the first kind of order unity, ν is a coordinate related to the lateral distance, r, in the focal plane. Where n is the refractive index of the medium in the space between the objective and the specimen, r is the real radial coordinate in image space, and λ is the wavelength of the incident light, the NA of the objective is n sin Θ, the coordinate ν is perpendicular to the optical axis and is a normalized coordinate:

$$\nu = \frac{2\pi}{\lambda} rn \sin\Theta. \tag{7.2}$$

For the confocal microscope case, a pinhole is in front of the detector (a point detector); the image intensity of a point source of light in the focal plane is now given by

$$I_{confocal}(\nu) = \left[\frac{2J_1(\nu)}{\nu}\right]^4. \tag{7.3}$$

These equations are for a point object and will be different for plane objects, as well as in the case of fluorescence (incoherent light). The key result is that the lens images a point object and an intensity distribution called the Airy disk. When we compare a confocal (point detection) with a conventional microscope, in which both image a point object in reflected light, we find that the Airy disk in the confocal microscope is narrower and the sidelobes from the central intensity peak are reduced, which is the physical explanation of the increased resolution obtained with a confocal microscope.

We now consider the axial distribution of light from a point source of light that is imaged with an aberration-free lens with a circular aperture. The light distribution is given by Eq. (7.4) for a conventional microscope:

$$I_{conventional}(u) = \left[\frac{\sin(u/4)}{(u/4)}\right]^2. \tag{7.4}$$

The normalized axial coordinate, u, is given by Eq. (7.5), where z is the real axial coordinate in image space, and λ is the wavelength of the incident light:

$$u = \frac{2\pi}{\lambda} z(\mathrm{NA})^2. \tag{7.5}$$

For comparison, we present the result for the confocal microscope:

$$I_{confocal}(u) = \left[\frac{\sin(u/4)}{(u/4)}\right]^4. \tag{7.6}$$

For the case of a point object, these equations show that for a confocal microscope the central peak is sharpened as compared to the conventional microscope.

The axial resolution is higher for a plane object than for a point object in reflection, and can be measured from a plot of the variation in the axial image intensity of a point object on the optic axis as the object is displaced through the focal plane. Experimentally, it is observed in a confocal microscope that the depth of field, which is operationally defined at the half-maximum height or a plot of intensity against axial distance, is reduced relative to a nonconfocal microscope by a factor of 1.4.

7.10 Comparison of Different Confocal Microscope Designs: Which One Should You Purchase?

How does the direct view, real-time, tandem-scanning microscope compare with the LSCM? How does beam scanning compare with stage or specimen scanning? The answers to these questions are related to the more general question: Which confocal microscope is best for my needs? The answer depends on what the user wishes to observe. A confocal microscope used to observe *in vivo* human skin or the living eye in the ophthalmology clinic will require different features than one used to study computer chips or integrated circuits. Different types of confocal microscopes may be optimal for intravital studies of the nervous system, embryo development, or single-molecule fluorescence studies. In the case of a microscopy center with multiple users, the choice of a confocal microscope would involve a compromise of several features. Confocal microscopes suitable for general use are never optimal for a specific use.

In general, commercial confocal microscopes are usually purchased and used. While the cost is high, there are many benefits; for example, the systems are ready to operate and the time from installation to use in research is short. There are also commercial courses that some manufacturers provide to compress the learning curve. The commercial confocal microscopes usually provide computer software to control the operation of the system, from changing lasers and filters to changing objectives.

As for which confocal microscope is best, the answer is simple: try before you buy. Either take your samples to a working confocal microscope in the area or arrange for the manufacturer to place a demonstration unit in your facility. What is critical is that you test your particular specimens over a period of time (weeks rather than days is preferred). Try all the modes and all the features that you think are critical to your research. That period of time is important to help you decide how the confocal microscope meets other requirements related to economic considerations, maintenance contracts, record of service for the user, number of independent channels required, possibility of conversion to a multiphoton excitation microscope in the future, upright or inverted configuration, and ease of modification, e.g., to add lifetime imaging. Most commercial systems have extensive specifications that are available. It is also important to speak with several individuals who use a particular commercial confocal microscope and hear about their experiences.

Another approach is to purchase the basic building blocks of a commercial confocal microscope and then modify the system to meet your specific needs. That option requires a broad knowledge of optics, mechanics, signal processing, computer interfacing, and system design. Still, the rewards can be great. It is possible to construct a confocal microscope with specifications unique to your requirements. Also, in the process of constructing your microscope, you experience the great learning process of doing it yourself. This approach can also be accomplished if skilled individuals are available to design the systems and machine and assemble the components. Today there are an abundance of precision mechanical, optical, and photonic components. Commercial software programs such as LabView® are useful for control functions, data acquisition, and manipulation. Also, there are a wide variety of computer software programs for image processing; for example, the free software package *ImageJ*, which is available for several computer platforms from the National Institutes of Health.

A third approach is to construct your own confocal microscope, by purchasing the laser light sources and the basic microscope, and adding commercial components to meet your requirements for scanning and detection. Some people design and construct custom circuit boards for control functions and data acquisition as well. There is a wonderful feeling of accomplishment when you design and construct your own confocal microscope, and it can be an educational experience for students and others who are involved in the process.

When determining what type of confocal microscope to buy, ask yourself what data acquisition rates are required for your studies. If you are only observing fixed, stained sections, then rapid image acquisition rates are unnecessary. If you are interested in observing rapid transient events, such as calcium spikes, or events in excitable tissue, then the kinetics of these events set higher requirements for image acquisition speeds. For reflected-light confocal microscopy of highly reflecting specimens such as hard tissue (bone, teeth) or semiconductor wafers, the real-time, direct-view tandem-scanning confocal microscope could be ideal. For other applications, a rapid beam scanning confocal microscope that uses acoustic-optical scanners may be ideal. Stage or specimen scanning confocal microscopes would not be suitable, as they are too slow.

Another important decision is whether to use stage scanning or beam scanning. Both types of confocal microscopes present different advantages and limitations. As previously stated, with stage or specimen scanning there are no off-axis aberrations to correct. In principle, the specimen could be quite large. If the specimen is stable, the stage could be programmed to scan the complete specimen even if the process required many hours. Beam scanning confocal microscopes scan the light beam over the back focal plane of the microscope objective. Therefore, many rays are not paraxial, and off-axis optical aberrations are a consideration. Image acquisition time in beam scanning is usually limited by the signal intensity at each spot or pixel in the scan. With a strong signal it is sufficient to have a short **dwell time** on each pixel to collect the light. With much weaker signals, i.e. low-level fluorescence, a longer dwell time on each pixel is usually required to obtain a sufficient signal –to noise ratio for good contrast and image quality. The disadvantage to increasing the dwell time is that the photobleaching rate increases and the viability of living specimens decreases. Alternatively, the aperture in front of the detector can be opened to increase the strength of the detected signal; but the wider aperture will adversely affect the optical sectioning capability of the confocal microscope.

It is important to match the microscope capabilities to the use of the instrument. Applications in medical research and biology generally use fluorescence techniques. In the transmission mode, it is possible to use DIC and other contrast methods such as fluorescence.

7.11 Limitations of the Confocal Microscope

What are the major limitations of tandem-scanning confocal microscopes and LSCMs? I have previously discussed the loss of light in the illumination side of a Nipkow disk-based confocal microscope. This loss of signal scales with the ratio of the aperture area to the disk area.

Another, more subtle loss of illumination occurs within the specimen itself. If the thickness of the specimen homogeneously scatters and/or absorbs the incident illumination light, then the intensity will be successively reduced layer by layer throughout the specimen. Alternatively, the specimen can have a highly absorbing or highly scattering layer on the surface or within its thickness. In each case, the illumination decreases at the lower layers of the specimen. Light from the illumination beam lost because of scattering or strong absorption cannot be regained and used to illuminate lower regions of the specimen. The observer will see a reduced signal intensity at the lower planes.

The main limitation is detection of the signal used to display the image. First, we ask, what limits the signal? On the illumination side, we can increase the brightness of the light. Typically, lasers are used as the excitation source. At first you may think that the brighter the light source, the stronger the induced fluorescence of the specimen, and therefore the stronger the detected signal. Can we increase the source brightness without limit? The answer is no.

The laser sources typically used with confocal microscopes are extremely bright, and the diffraction-limited focused spot on the specimen is many times the brightness of the surface of the sun. If the specimen to be observed is living—for example, any specimen of live cells, live tissue such as brain slices, a developing embryo, or clinical *in vivo* microscopy of human eyes or skin—then there is a chance for light-induced damage to the specimen, which could be caused by thermal, mechanical, photochemical, photophysical, or any combination of these means. The goal of intravital microscopy is to minimize damage to the specimen induced by illumination. The same goals hold for human clinical microscopy, for which there are added safety and ethical considerations. Light-sensitive organs such as the eye are only permitted to be subjected to light levels that have been shown to be safe.

The second process that limits the useful intensity of the illumination light is the photophysics of the fluorescence process. In the absence of light, the fluorescent molecules (whether naturally occurring, such as NAD(P)H or serotonin, molecules labeled with fluorescent probes, or genetically expressed fluorescent molecules) are in their ground electronic state. Only when illuminated with the appropriate light are they excited to higher electronic states. When the fluorescent molecules return to the ground electronic state they emit light, which is what we call fluorescence. The process of fluorescence is not instantaneous; it has a finite duration, which is measured by the fluorescence lifetime (typically nanoseconds). As we increase the intensity (brightness) of the illumination light from zero to increasing intensities, we first observe an increase in the fluorescence intensity. Above a certain threshold, which is a property of a particular fluorescent molecule and its electronic structure and energy states, we observe that the fluorescence intensity saturates. It is the **fluorescence saturation** of a population of fluorescent molecules that places an upper limit on the intensity that we use to induce the fluorescence. The physical basis for the fluorescence saturation is that at high light intensities, all of the molecules are in the excited state, and there can be no further absorption of exciting light. Only when a molecule emits a photon and returns to the ground state can absorption recur.

Once the specimen is induced to emit fluorescent light, the goal is to collect and detect every photon from the focal plane. Since we are talking about both confocal microscopes with their conjugate apertures to discriminate against defocused light, we limit our arguments to the focal plane. Assume the fluorescent molecules are randomly distributed in the focal plane; that means the molecules have an equal probability of emitting a photon in the fluorescence process in any direction. If we further assume that these emitted photons all escape the specimen (although there could be multiple points of scattering along the escape path), then we see the immediate problem.

Since every photon contains information, every photon counts and should be collected, detected, and contribute to the displayed specimen image. When the intensity of the emitted fluorescence is high with respect to other losses in the microscope and noise and quantum effects in the light detectors, the loss of many photons is not very important. However, when the intensity of the detected fluorescence is very low with respect to problems of detecting these few photons, then we are up against real limitations. In order to collect all these photons, the microscope objective should completely surround the specimen! As many photons will be emitted in the direction of the microscope objective as will be emitted in the opposite direction, and are therefore undetected.

Assume we use high-NA microscope objectives with the correct thickness of cover glass and immersion fluid for the objective; then only those emitted photons that enter the acceptance angle of the microscope objective (higher NAs collect more light) can be detected.

Between the microscope objective and the light detector, considerable light loss occurs in the confocal microscope. In general, each optical element, whether mirrors, filters, polarizers, or lenses, contributes to the loss of light. As the number of optical surfaces increases, the loss of light increases, which is assuming the use of state-of-the-art antireflection coatings on each surface. Therefore, the goal is to design confocal microscopes with a minimum number of surfaces that result in light losses.

So, the remaining light that comes from the specimen is to be detected. Note that in low-level fluorescent confocal microscopy, electronic detectors typically are used. With the direct-view tandem-scanning confocal microscope it is possible to use a two-dimensional imaging device such as a film or charge-coupled device (CCD) camera. LSCMs usually use a low-noise photomultiplier to detect the fluorescence light.

For extremely low levels of fluorescence at the face of the detectors, the quantized nature of light (discrete photons) and the sources of noise in the detector and the associated amplifiers place limitations on how few photons or how low the light intensity can be and still be detected.

When our eyes are correctly adapted, they can detect single photons over a range of several log units of intensity. Solid state light detectors are less efficient. At best, we may achieve 90% detection efficiency with CCD cameras, but typically the value of the quantum efficiency is less than that. The **quantum efficiency** is the percent of photons incident on the detector that generate a signal. For a photomultiplier the quantum efficiency, which depends on wavelength, may be in the range of 1% to 40%. Note also that the detection efficiency of solid state detectors also varies with wavelength. Usually, these devices are selected to have high quantum efficiency in the wavelength regions that match our experimental conditions.

An important consideration is the statistical distribution of detected photons. The quantized nature of photons in a light beam is the physical cause of the distribution. For a coherent light source this probability distribution can be described by the Poisson distribution, which is valid for an ideal laser source that emits monochromatic light; the light is coherent and single-mode. As an example, if 100 photons are generated in a beam of light, then the inaccuracy in detection is about $\pm$ 10 photons.

Our ideal laser is assumed to output photons in a random manner. If the mean value of the average intensity is n, then the noise associated with this average signal is given by the root mean square of n. A useful parameter is the SNR. Typically, the SNR is equal to the mean value squared, divided by the variance. For the Poisson distribution, the SNR is equal to the mean value of the signal. This has an important consequence: the SNR increases as the mean number of photons, or signal strength, increases.

Another set of problems limits the use of fluorescent confocal microscopes for intravital microscopy of thick, highly scattering specimens. Confocal microscopes, their light sources, their microscope objectives, and all the optical components are designed, optimized, and constructed for use with visible light. When these microscopes are adapted with ultraviolet light sources without extensive modification of the components, new problems arise in three classes: (1) the requirement to drastically modify confocal microscopes that were optimized for visible wavelengths; (2) ultraviolet damage to live cells, tissues, organs and embryos; and (3) the limited penetration depth of ultraviolet light due to enhanced light scattering in thick specimens.

First, many examples of fluorescence confocal microscopy exist in which the fluorescence is excited in the ultraviolet wavelength region. For example, studies of NAD(P)H fluorescence in cell metabolism, measurements of intracellular ion concentrations with ion-indicators that absorb in the ultraviolet, photoactivation or uncaging of trapped ions and molecules that are triggered with ultraviolet light, studies of cellular autofluorescence, and fluorescence of many biological molecules with absorption peaks in the ultraviolet. Several investigators modified commercial confocal microscopes by introducing lasers and other light sources with ultraviolet light; furthermore, they changed the microscope objectives, beamsplitters, filters, optical coatings, and other components to maximize the system for use with ultraviolet light. While these drastic modifications had partial success in individual instruments, other limitations arose.

It was demonstrated many years ago that ultraviolet light damages and is lethal to living cells, tissues, and organisms. The use of a confocal microscope with highly focused ultraviolet light is dangerous for *in vivo* human studies such as diagnostics in ophthalmology and dermatology. When used for fluorescent intravital microscopy over an extended period of time, the high-energy ultraviolet light is absorbed by critical cellular components and induces damage and eventual death. This precludes the use of ultraviolet confocal microscopy to study the development of embryos and long-term observation of cells in tissue culture.

Also, it is observed that the penetration depth of ultraviolet light into thick biological tissues is less than that with longer-wavelength visible light. That effect results from increased light scatter and increased absorption of the ultraviolet light in cells and tissues.

All these limitations are severe and place limits on biological studies that require the advantages of a fluorescent confocal microscope but with the added constraint of ultraviolet excitation of the fluorescent molecules of interest.

All confocal microscopes have limitations. The signal strength can be degraded by multiple sources: illumination efficiency of Nipkow disk-based confocal microscopes; absorption and scattering of the illumination light within the upper regions of the specimen, which reduce the illumination intensity at the lower regions of the specimen; reflecting surfaces within the microscope resulting from multiple optical components and surfaces; stray light reaching the detector; misalignment of the components; dirt and dust on the apertures and other optical surfaces; quantum efficiency of the detector; the quantized nature of light; the signal to noise ratio; the image contrast; and the type of object imaged.

The partial solution to these formidable problems came with the development of nonlinear microscopy, specifically multiphoton excitation microscopy. Part III is devoted to this important advance.

7.12 Summary

- The greatest advantage of the confocal microscope is the elimination of out-of-focus light (depth discrimination). Point illumination and conjugate point detection is the principle, with the images of both apertures cofocused in the specimen. Spatial filtering is used to eliminate the out-of-focus light. Optical sections of less than 1 μm thickness can be imaged within thick, scattering tissue.
- The optical sectioning (depth discrimination) capability of a confocal microscope is the basis of three-dimensional microscopy. With defocus, the image becomes darker and disappears. A conventional fluorescence microscope has no optical sectioning capability; with defocus, the signal is constant but fuzzier.
- Consideration of signal intensity and the related signal-to-noise ratio is of paramount importance for detection.
- Marvin Minsky is credited with the experimental realization of a stage scanning confocal microscope. He clearly stated the advantages of stage or specimen scanning in his 1961 patent on the confocal microscope. This idea decoupled the magnifications of the objective from the resolution. The magnification could be changed by changing the number of pixels in the image. His patent also clearly showed the folded (reflected) mode of modern confocal microscopes.
- Minsky pointed out an important advantage of specimen or stage scanning. The microscope only used the central part (paraxial optics) of the microscope objectives, thus there are no off-axis or lateral optical aberrations to correct. Chromatic and spherical aberrations still required correction.
- Minsky's confocal microscope used a 45× microscope objective in air. It could resolve points closer than 1 μm apart. Its disadvantage was the slow scan time: 1 frame per 10 seconds.
- A real-time tandem-scanning confocal microscope, in which the image could be observed with the naked eye, was developed by Petràn and Hadravsky. They

decided to use a multiple-aperture (multibeam scanning) device since it would reduce the frame time to scan the field as compared to single-point scanning.

- The advantages of the Nipkow disk-type confocal microscope include real-time viewing, true color observation of the specimen using color to map the depth of the features in the specimen when used with an objective with low correction of chromatic aberrations, and direct-view observation.
- The tandem-scanning Nipkow disk-based confocal microscope is a poor choice for weakly reflecting specimens such as living cells, tissues, and organs. Consequently, it is not suitable for imaging weak autofluorescence or weakly stained fluorescent specimens. This is because of the low illumination efficiency of the Nipkow disk.
- Xiao, Corle, and Kino invented a real-time, one-sided, Nipkow disk-based confocal microscope. This design has several advantages over the tandem-scanning confocal microscope: it is less sensitive to vibration of disk, has a simplified optical design, and is easier to align.
- In the Yokogawa Nipkow disk confocal microscope, a laser illuminates the upper spinning disk, which contains about 20,000 microlenses over the pinholes. The lower disk contains pinholes in the focal plane of the microlenses that are arranged in the same pattern as the microlenses on the upper disk. Both disks rotate on a common axis. There is a great increase in light throughput.
- The key development in the Svishchev invention of a confocal light microscope was a scanning system based on an oscillating two-sided mirror. The two-sided mirror scans and descans a slit of light on the specimen side and on the image side.
- In a confocal microscope, if the illumination light is confined to a diffraction-limited spot on the specimen by the microscope objective, and the detection is also confined to the same spot with a pinhole aperture placed in front of the detector, then the confocal microscope would strongly discriminate against light from above and below the focal plane. The detector pinhole aperture is in a plane conjugate with the plane containing the illumination spot. With this confocal arrangement, the intensity of a point source of light on the specimen falls off with the fourth power of distance from the focal plane.
- A modern LSCM will only perform optimally if the following conditions are met: the system should be mounted on an anti-vibration optical table; laser-cooling fans and other sources of vibration should not be placed on the optical table; the optical elements should be correctly aligned and free of dirt, oil, and scratches, especially the microscope objectives.

Chapter 8

The Development of Scanning-Slit Confocal Systems for Imaging Live Cells, Tissues, and Organs

Chapter 8 will describe and compare a number of confocal microscopes based on slits instead of pinhole or iris apertures. As with conjugate pinholes, the images of two slits (one on the illumination side and one on the detection side) are cofocused on the specimen. A slit has width and length, and these dimensions are different; a pinhole aperture is circularly symmetric. The most important difference is that the axial and transverse resolution will be different in the direction of the slit width and in the perpendicular direction. What is gained in scanning-slit confocal microscopes is a decrease in image acquisition time, since the slit image scans across the specimen, which is equivalent to scanning many points simultaneously (all points along the length of the slit image are scanned in parallel).

The use of slits in a Nipkow disk-based confocal microscope is the basis of the Lichtman and Sunderland invention of a new confocal microscope (see his patent in Masters, 1996). This microscope was first developed for imaging acetylcholine receptors on muscle-cell membranes. Because it is conceptually different from the scanning-slit confocal microscopes described in this chapter, it is not discussed further.

With the exception of Svishchev, who was motivated to image the live brain cortex, and Baer, who wished to develop a microscope for biological imaging of live cells and tissues, the group of inventors discussed in this chapter were interested in imaging live cells in the living eye. The living eye presents unique problems for *in vivo* light microscopy; its movement is the principle problem. Specifically they were interested in developing optical microscopes that could image live cells in the unstained cornea. While the initial inventions solved some problems associated with imaging the *ex vivo* cornea, the progression towards imaging the human cornea *in vivo* continued until it became a clinical reality.

This chapter presents a series of linked technical advances in the development of scanning-slit confocal microscopes, with applications in the fields of biology and ophthalmology. Each technical advance was the partial solution to a problem. The end result is the development of a scanning-slit clinical confocal microscope to examine the human eye *in vivo*.

In this historical context several lessons can be learned. First, a clear statement of the problem is necessary for its eventual solution. Second, curiosity is a strong motivator for technical innovation. Third, multidisciplinary research and technical development often results in advances that would not occur within a single discipline. Finally, advances in the biomedical field often involve the synergistic combination of medically trained individuals and physicists and engineers.

8.1 Scanning-Slit Confocal Microscope

To start, it is necessary to explain some terms and their different uses by the individual inventors and authors. A **slit aperture** consists of two blades or plates that can be synchronously displaced from the state in which the plates are touching each other and no light is transmitted, to the state in which the plates are separated and light can be transmitted through the rectangular opening. When the slit opening can be varied, the slit is termed an **adjustable slit**. When the rectangular opening is fixed, for example when the opening is machined or etched into a plate or formed by evaporating metal over a mask on a quartz plate, then the device is termed a **fixed slit**. The opening of a slit is termed the **slit width**. In an adjustable slit the opening or the slit width can be changed. The long dimension of the slit is termed the **slit length**.

In this chapter the authors of the various scanning slit confocal microscope have used the terms wide and narrow slits and wide-field microscope. In their papers and patents, the meanings of these terms differ from definitions in the Royal Microscopical Society *Handbook of Light Microscopy*. For example, the slit width of the adjustable slits is sometimes stated as "wide" or "narrow." In this context, a wide slit has a large width or opening, and a narrow slit has a small width or opening. A wide slit can pass more light than a narrow slit.

The second term is "wide field," which refers to the size of the field of the endothelial cells that are imaged. When only a few endothelial cells are imaged, then the image is termed a "narrow field of view." When many endothelial cells constitute the image, the authors termed the image a "wide field of endothelial cells." Charles Koester called his confocal microscope "a wide-field specular" microscope, which was termed relative to the early corneal microscopes that could only image a few endothelial cells at a time, i.e., 6 to 10 cells; Koester's microscope resulted in an image of many thousands of endothelial cells. As will be explained in subsequent sections, the number of endothelial cells in the image was constrained by the use of slits of very narrow width, which was required in order to provide a narrow depth of focus needed to produce images with sharp cell borders and high contrast. Alternatively, when the microscope was used with a large slit width (wide slits), the larger depth of field caused the image of the endothelial cell layer to become blurred and of low contrast. Note that this definition of wide-field is different from when it refers to a microscope in which the illumination system illuminates multiple regions of the specimen in parallel.

As you will see in the following section, the use of variable slits in a scanning-slit confocal microscope has great utility. Both the NA of the microscope objective and the width of the confocal slits determine the thickness of the optical section. To achieve a thin optical section with a given microscope objective, which means an increased rejection of scattered light from adjacent sections, it is necessary to make the slits very narrow. That also results in a reduced signal at the detector.

On the other hand, when the signal is limited, it may be necessary to open the slits. Wider slits will result in both an increased signal and a decreased rejection of out-of-focus light in the image. Depending on the specimen, the user will select a slit width that is a compromise between optimal optical sectioning and signal strength. These arguments are appropriate for the scanning-slit microscopes developed by Svishchev, Maurice, Koester, Masters, and Thaer.

What is a scanning-slit confocal microscope and why is it useful? One may think of a slit as a linear array of pinholes. While a pinhole can be described in terms of its diameter or radius, a slit is described by its length and width. In general, scanning-slit systems are useful in cases when rapid scanning is important, and also when it is necessary to collect more light from the specimen than is possible with pinhole-based scanning systems. A scanning slit can scan the equivalent of many pinholes in parallel, and that results in its rapid scanning compared with point-scanning systems.

Scanning-slit confocal microscopes have the following advantages over point-scanning systems: (1) more light illuminates the sample and enters the detector; (2) the slit width is easily adjustable to compensate for different amounts of light scattering within the sample; (3) the design and construction of scanning-slit systems is relatively simple.

Nevertheless, the specific types of scanning-slit confocal microscopes that are discussed in this chapter also have several disadvantages: (1) the optical sectioning (depth discrimination) capability is different along the length and the width of the slit, (2) a wide slit will not perform as well as small pinholes with respect to optical sectioning capability; (3) the full NA of the microscope objective is not utilized; rather, one-half of the NA is used for the illumination path, and the other half to collect reflected light; (4) several implementations of scanning-slit confocal microscopes flatten or applanate the specimen, which results in alteration of its microscopic structure. Some scanning-slit confocal microscopes do utilize the full NA of the objective, such as that invented by Burns et al. (Burns et al., 1990).

In the examples given below, the driving force was to image the eye *in vivo*. These objectives had the following constraints: (1) the microscope must not result in eye injury; (2) the specimen is a weakly reflecting, semitransparent object with low contrast; and (3) the living eye may move. The inventions of Svishchev and Baer, while not motivated by the desire to image the eye, are included in this chapter because their work was appropriated by subsequent inventors and had a strong influence on the developments of clinical *in vivo* microscopy of both the eye and the skin.

8.2 Statement of the Problem: Slit Width Versus Field of View

The following sections trace the development of a clinical ocular instrument that could be used to observe the thick tissues of the living eye with a thin optical section and a wide field of view. The requirement of a thin optical section results in high-contrast images within the focal plane of the confocal microscope. The requirement of a wide field of view results in an image that spans many square microns and thus permits observation of many cells within the field of view.

8.3 Goldmann's Wide-Field Microscope

Hans Goldmann, an ophthalmologist at the University of Bern, Switzerland, developed in 1940 a confocal instrument for slit-lamp photography of the anterior segment (cornea and ocular lens) in which the entire optical section could be imaged sharply onto the film, thereby increasing the depth of field of the instrument. Goldmann's modification of the Gullstrand slit lamp used a photographic system that moved on the optical axis (see Fig. 8.1). Both the camera and slit beam mechanically moved forward during the exposure, and the film traversed synchronously behind the slit; therefore, the image on the film is maintained in continuous focus. This results in a sharp image of the entire optical section and solves the problem of a limited depth of field. Goldmann's innovative technique included a system that could integrate the images from the various adjacent optical sections into a composite image that was in focus across the entire thickness and showed high contrast. This concept was the basis of the future works of David Maurice (wide-field specular microscope), Charles Koester (wide-field specular microscope for *in vivo* use), and Andy Thaer (scanning-slit clinical confocal microscope).

8.4 Maurice Invents Several Types of Specular Microscopes

The cornea is the anterior region of the eye. It is semitransparent and surrounded with fluid on both sides: on the anterior surface there is a thin tear film, and on the posterior surface is the aqueous humor. The human cornea has a central thickness of approximately 520 μm. The structure of the cornea from the tear side to the aqueous humor side is as follows: the epithelium consists of about five layers of epithelial cells and is ~ 50 μm thick; Bowman's layer is an acellular layer between the epithelium and the stroma, where the stroma consists of keratocyte cells and orthogonal arrays of collagen fibers and is ~ 460 μm thick; Descemet's membrane is an acellular membrane between the stroma and the endothelium; and, finally, the most posterior limiting layer is a single layer of endothelial cells approximately 5 μm thick.

Differences in refractive index between the cornea and the tear film and between the endothelium and the aqueous humor result in strong specular reflections. These reflections can be used to image these limiting cell layers. In the 1930s, Vogt modified the slit lamp of Gullstand in order to image the specular reflection of the

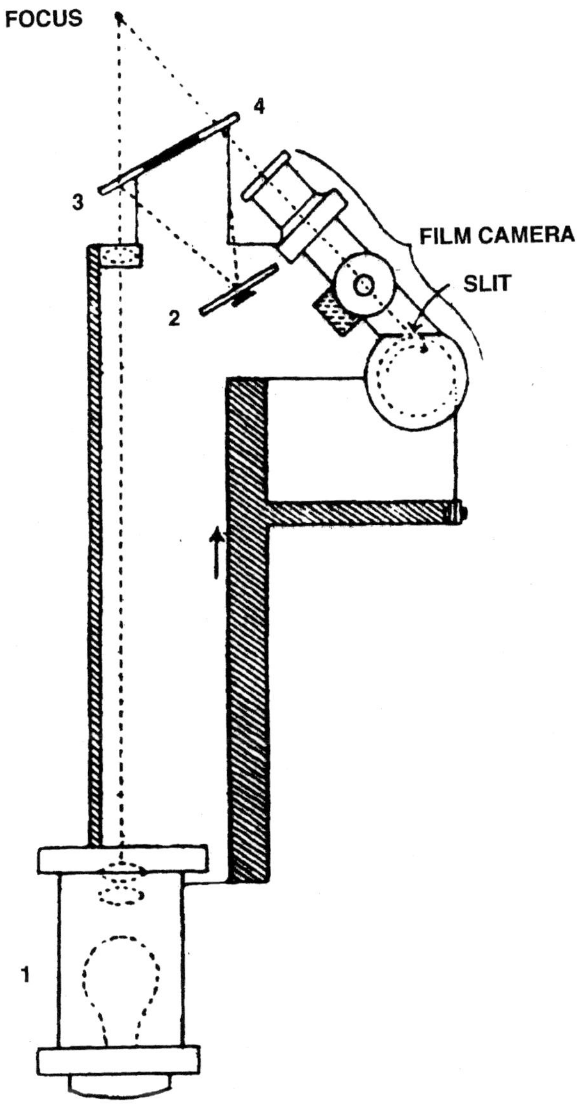

Figure 8.1 Schematic diagram of the instrument developed by Goldmann for slit-lamp photography and photometry.

corneal endothelium at the corneal–aqueous humor interface, which acts as a mirror during focal illumination. Goldmann used focal point illumination and the specular reflection from the corneal endothelium to observe endothelial cell patterns at low magnification. This is an amazing feat. The optical instrument passes a narrow slit of light through the 500 μm of the cornea and collects the specularly reflected light from the 5-μm-thick corneal endothelium at the posterior corneal–aqueous humor interface. The large difference in refractive index at this interface results in the high reflectivity of the specular reflection image.

While many types of optical microscopes were developed to view semi-transparent specimens, another class of opaque specimens required novel designs. In 1910 the Nachet Optical Instrument Company published a microscope catalog that included a metallurgical microscope. This instrument was designed to view opaque, polished metallurgical objects. The key features were illumination from a vertical tube positioned above the microscope objective and perpendicular to the optical axis of the microscope. This type of illumination system is generally called a vertical illumination system (see Fig. 8.2). One half of the microscope objective was used for illumination of the sample, and the other half was used to collect the reflected light (see Fig. 8.3). While this optical arrangement only uses one half of the microscope objective NA, which results in less resolution, it offers good separation of the illumination light from the reflected and scattered light. A small prism or an angled mirror was used to deflect the horizontal illumination light downward toward the specimen.

Maurice developed a high-magnification (500×) specular microscope to photograph cells from an *ex vivo* eye (Maurice 1968). Vogt's term "Spiegelmikroskopie" was translated into English as "specular microscopy" and used initially by Maurice. Maurice's specular microscope with a 40× water-immersion microscope objective, focal slit illumination with a yellow or orange filter to improve contrast, and an applanating cover glass to flatten the cornea, is shown in Fig. 8.2. The aperture of the microscope objective is divided across its center. A mirror at an angle directs the illumination light down one side of the objective and the reflected light

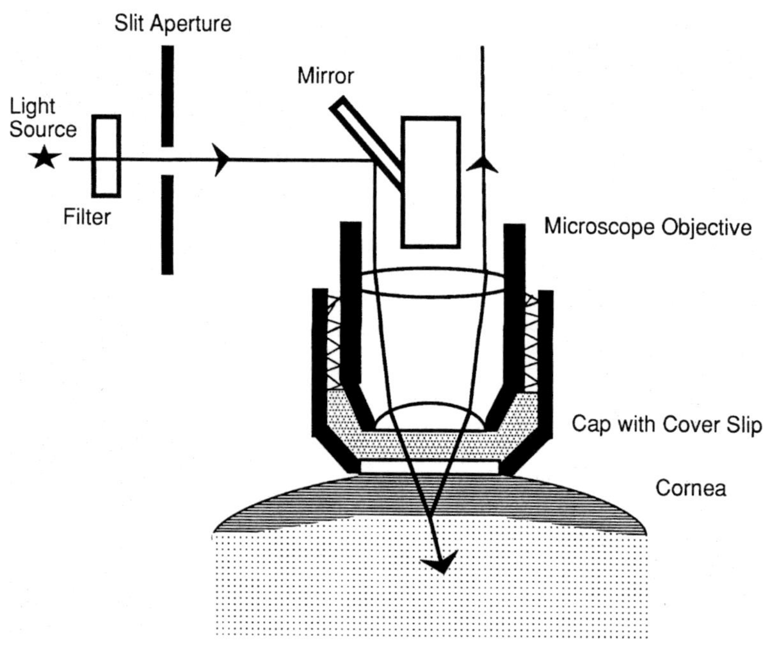

Figure 8.2 Schematic diagram of the Maurice specular microscope.

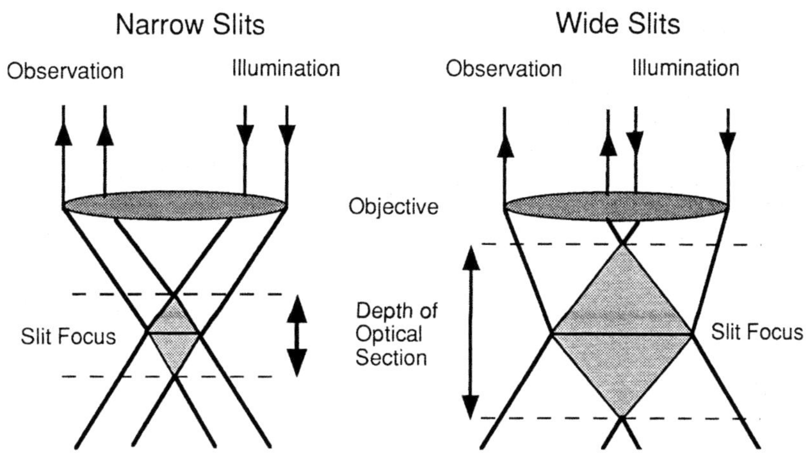

Figure 8.3 Schematic diagram of the optical principle of the Maurice optical microscope under conditions with narrow slits (left diagram) and with wide slits (right diagram).

from the endothelium–aqueous interface passes up the opposite side of the objective. The similarities with the previously described metallurgical microscope are obvious. Maurice once told this author that the design of the early metallurgical microscope gave him the idea for the illumination in his first specular microscope. The resulting high-magnification image of the corneal endothelium can be viewed with the naked eye in the ocular or photographed. A cover glass fixed to a cap that screwed to the outside of the microscope objective is used to applanate (flatten) the cornea and improve its stability. The light source is a slit of illumination focused in the focal plane of the microscope objective.

Maurice subsequently modified the scope by placing the fine adjustment on the microscope instead of the cap on the microscope objective. Maurice reported imaging the epithelium, stroma, and endothelium at 500× magnification and noted that both the corneal thickness and the thickness of the individual layer could be measured to within 2 μm. Another change was that the *ex vivo* eye was not applanated. A water-immersion microscope objective with a working distance of 1.6 mm and 40× power was used, and saline covered the cornea and separated the microscope objective from the cornea. Maurice published images of the endothelium of a young and an old donor eye at a magnification of 500×.

The contributions of Maurice at this stage of his invention include use of a high-NA water-immersion microscope objective, pressing the microscope objective onto the eye to flatten the cornea, alternatively immersing the eye in saline and using the microscope objective in a noncontact mode in the saline solution, use of a split microscope objective with one half for illumination and one half for light collection, and the use of a narrow slit of light for illumination. The problems still unsolved were how to observe a narrow optical section and a wide field of view, and how to observe *in vivo* eyes without motion artifacts and blur. About the same time, in another corner of the world, a scientist was inventing a confocal microscope to study living cell and tissues in the brain.

8.5 Svishchev's Invention of a Scanning-Slit Confocal Microscope

Driven by the need to investigate neural tissue in the *in vivo* brain, Svishchev invented a scanning-slit confocal microscope. The key development was a scanning system based on an oscillating two-sided mirror. This design eliminated the need for precision-controlled galvanometer mirrors for the scanning and descanning; a simple two-sided mirror mounted on an oscillating rod performs the synchronized functions of scanning, descanning, and rescanning for viewing. The details of the scanning-slit microscope were described in Sec. 7.7 and illustrated in Fig. 7.7.

The oscillating two-sided mirror in the Svishchev confocal microscope was subsequently incorporated into the Thaer clinical microscope.

8.6 Baer Invents a Tandem-Scanning-Slit Confocal Microscope with an Oscillating Moving Mirror-Slit Assembly

In 1970, while a Ph.D. student at the Albert Einstein School of Medicine, New York, Baer invented a novel confocal microscope to examine biological specimens at high resolution and contrast. His confocal microscope was characterized by the following features: a divided-aperture microscope objective with half the objective used for illumination and half for viewing; and a lightweight, oscillating, rigid assembly consisting of two conjugate (confocal) fixed-width slits and a hemispherical mirror for scanning and descanning (see Fig. 8.4). Baer provided an elegant device in which both the illumination and the collection slits were mounted on a fixed assembly that undergoes rapid oscillation, thus ensuring that both slits scan in phase. It used focal-plane-specific illumination; the light from the out-of-focus plane is selectively masked from the light detector or the eyepiece.

Baer suggested the use of broadband incoherent light sources that are suitable for use with a wide selection of fluorescence probes for staining. The images of the biological specimens are in true color. In his patent, Baer suggests two important techniques for three-dimensional microscopy: (1) an objective lens with deliberate longitudinal chromatic aberration that will result in color-coded depth scans formed with a one-dimensional slit scan; and (2) chromatic dispersing prisms in front of the eyepieces that can produce real-time stereo imaging.

Baer thus provided another simple, elegant invention to obtain a wide field of view and a narrow optical section. Baer was aware of the previous work of Svishchev and cited the prior work in his 1979 Ph.D. thesis. The slits on the oscillating "T assembly" were of fixed slit width, thus losing the advantage of the adjustable set of slits in the Svishchev confocal microscope.

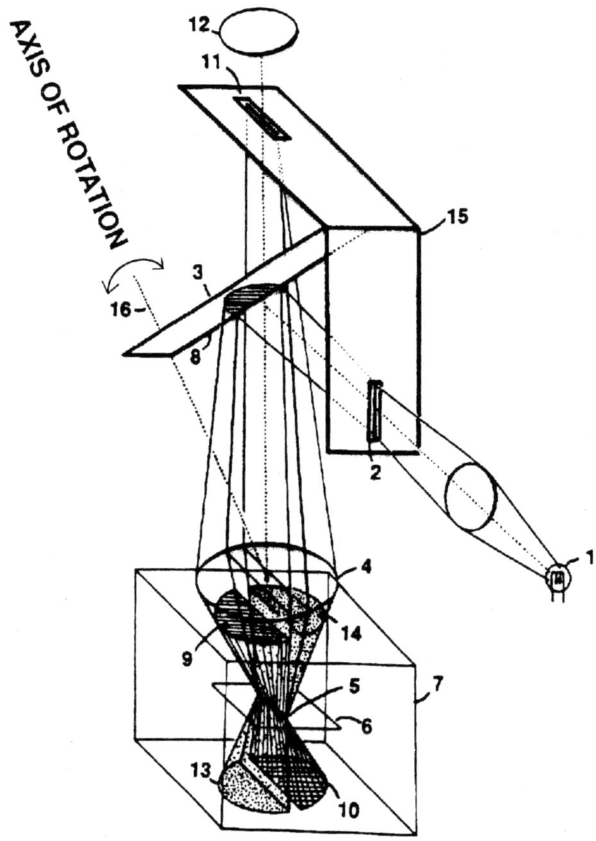

Figure 8.4 Schematic diagram of the Baer tandem-scanning-slit confocal microscope with an oscillating moving mirror-slit assembly. The 'T'-shaped assembly, 15, contains the illumination slit, 2, the hemispherical mirror, 3, and the second confocal slit, 11, oscillated about the axis, 16. Light from the source, 1, is imaged through the slit, 2, and is directed by the mirror, 3, to the back focal plane of the objective, 4. One side of the objective is used for illumination, and the other side of the objective is used to collect the scattered and reflected light from the sample, which is passed through the viewing slit, 11, and is imaged by the lens, 12, onto the detector (eye or film). The two conjugate slits form an image of a thin optical section, 6, that excludes scattered light from above and below the focal plane. The illuminated regions, 9 and 10, intersect the viewable regions, 13 and 14, only at the illuminated strip, 5. Part 7 is the specimen.

8.7 Maurice Invents a Scanning-Slit Wide-Field Specular Microscope

The problem inherent in Maurice's first microscope was that, when the conjugate slits were wide, the image details were obscured by out-of-focus scattered light from the surrounding tissue. When the conjugate slits were very narrow, the light scatter from sections adjacent to the focal plane of the microscope objective is much reduced, but the field of view is very small. This trade-off is illustrated in

Fig. 8.3, which demonstrates the effect of wide and narrow slits on the depth of the optical section and the field of view.

The scanning-slit optical microscope developed by Maurice solved this problem by using a narrow slit, slowly moving the object through the focal slit and recording the image on a film that undergoes a synchronized movement with the object (see Fig. 8.5). The optical performance of the microscope was remarkable: the z-axis (axial) resolution was 3 μm.

Maurice wrote that Goldmann's invention made use of a similar technique for photographing the cornea. His paper cited references to both the Baer U.S. patent 3,547,512 and the 1940 Goldmann paper, and stated that the superior feature of the Baer invention was that it could be used to directly view the cornea.

This new instrument used two stationary, conjugate, narrow slits, one forming a slit image of the light source, and one in a conjugate plane in front of the ocular. A wide field of view was achieved by synchronous translation of the specimen (an *ex*

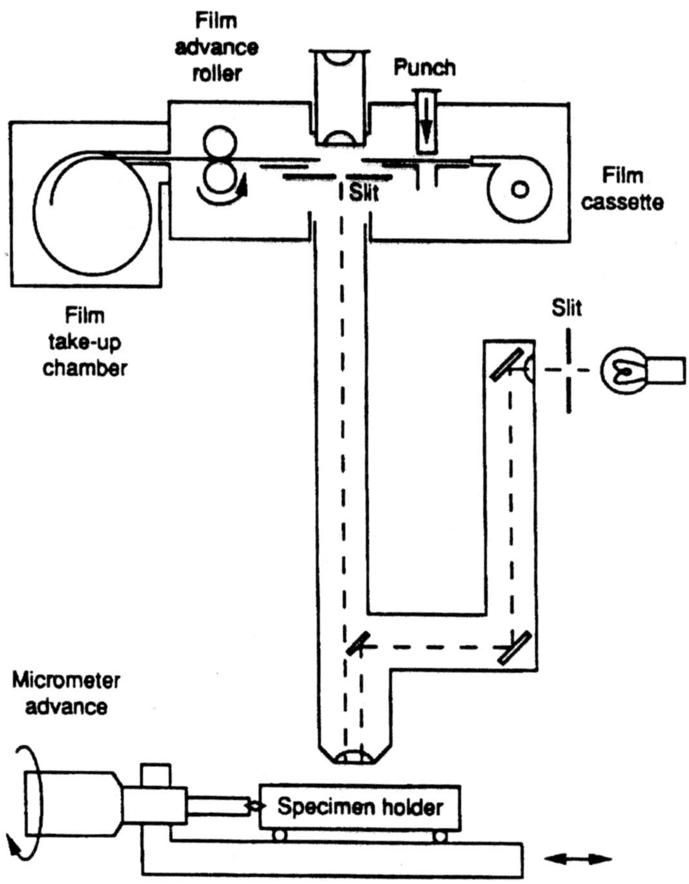

Figure 8.5 Schematic diagram of the Maurice scanning-slit optical microscope. Both the film advance rollers and the specimen holder advance synchronously in opposite directions when driven by identical synchronous motors.

vivo eye) and the photographic film. A high-NA water-immersion microscope was used, with half the NA used for illumination and the other half to collect the scattered light from the specimen.

While Maurice solved several of the problems associated with confocal microscopy of thick, highly scattering specimens, one significant problem remained: the slow, synchronous scanning of the specimen and the film precluded *in vivo* imaging.

8.8 Koester Invents a Wide-Field Confocal (Specular) Microscope for *In Vivo* Imaging

When the slits on the specular microscope were adjusted to a small slit width, the thickness of the optical section was reduced, but the field of view was also very small. In 1980 Koester solved this problem with the use of an oscillating three- or four-sided quartz prism or cube with metallized, highly reflective surfaces.

The oscillating prism or cube performs three functions: to scan, descan, and rescan the light for viewing. The light from the first slit is scanned over the specimen by the first facet of the oscillating mirror cube. The reflected light from the specimen is descanned by the second facet; therefore, the light is stationary at the second conjugate slit. Finally, the light from the second slit is rescanned by the third facet, which allows viewing by a stationary detector, i.e., the eye, a film camera, or a two-dimensional electronic detector (see Fig. 8.6).

In the Koester confocal microscope are two adjustable slits, located in conjugate planes. The principle is to scan a strip of illumination across the object plane and to

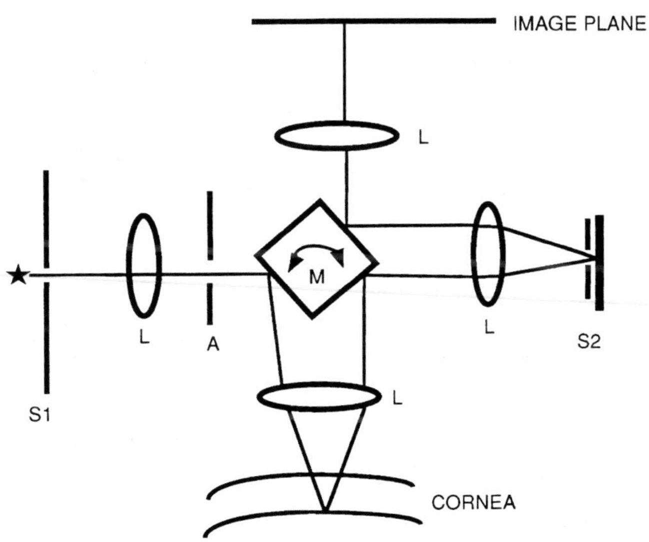

Figure 8.6 Schematic diagram of the Koester scanning mirror microscope. A light source (star) illuminates the first slit, S1, and one face of an oscillating mirror, M, then is reflected into one side of a microscope objective, L, which focuses the light in the cornea.

synchronously scan a slit-shaped viewing aperture across the same object plane, and to image the viewing aperture in the ocular image plane or film plane. The second confocal slit serves to select only that portion of the object plane that is illuminated by the first slit and to eliminate or mask out all other light sources from other regions within the specimen. While the microscope was designed for eye examination, Koester also proposed to use it to observe other biological tissues and organs such as the cochlea. The Koester patent also includes a drawing of a scanning microscope apparatus based on an oscillating two-sided mirror; this design is identical to the invention of Svishchev published in *Optics and Spectroscopy* in 1969. The Koester patent was filed in 1979, 10 years after the Svishchev invention was published in English. In his 1980 patent Koester cited the 1972 patent of Baer of a LSCM based on two confocal pinhole apertures and a set of mirrors for beam scanning.

The large number of reflecting surfaces in the Koester confocal microscope resulted in significant loss of signal. To compensate, the microscope was usually operated with the slits opened wide. The use of the wide slits together with half of the microscope objective NA degraded the resolution and the depth of the optical section within the cornea more than was optimal. Later versions of the microscope improved this with the development of custom microscope objectives with a higher NA, and the use of highly sensitive CCD cameras. These changes, which improved efficiency in both illumination and detection, permitted the two conjugate slits to be adjusted to a small slit width with a concomitant improvement of the thickness of the optical section.

A second disadvantage of his invention was that the custom-designed microscope objectives were in direct contact. His confocal microscope is comprised of two conjugate adjustable slits, an oscillating multisided mirror, a divided-aperture microscope objective that used one half of the NA for illumination of the specimen and one half for collection of the scattered light. An aperture with a central divider strip separated the illumination light from the reflected light on the side of the object. Since the full NA was not used, there was a loss of resolution.

The applanating microscope objective was custom made and not interchangeable with commercial microscope objectives. In addition, the applanating objective flattened the cornea and induced artifacts (i.e., dark bands) in the corneal images.

In spite of these limitations, Koester's confocal specular microscope offered a partial solution to the problem of *in vivo* confocal imaging of the cornea.

8.9 Masters Develops a Confocal Microscope based on the Maurice Design with an Axial Scanning Microscope Objective

In order to investigate the effects of contact lenses on the oxygen concentration and epithelial redox state of an *in vivo* rabbit cornea, it was necessary to design a non-imaging microscope with depth discrimination. In 1988 Barry Masters modified the specular microscope designed by Maurice to include a confocal redox fluorometer, which measured the NAD(P)H fluorescence and the scattered and reflected light along the optical axis of the *in vivo* cornea.

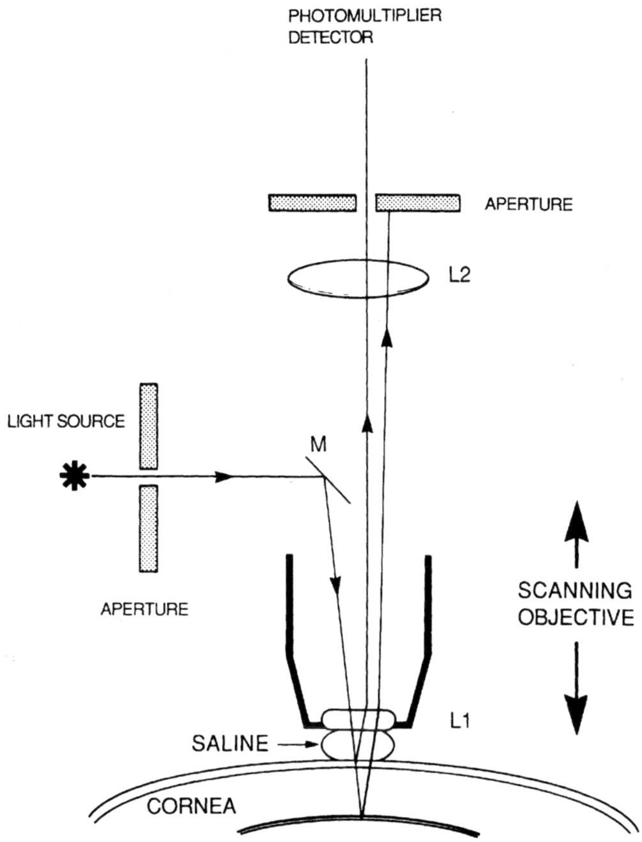

Figure 8.7 Schematic diagram of the axial scanning objective confocal microscope developed by Masters based on a modification of the Maurice specular microscope. The mirror, M, reflects light into one side of a high NA water-immersion microscope objective; scattered and reflected light is collected by the other side of the objective and passes through a second slit aperture conjugate to the first.

Masters addressed the problem of how to rapidly scan the *z* axis (thickness of a specimen) with an *in vivo* confocal microscope (see Fig. 8.7). First, the microscope features a 50×, NA 1.0, water-immersion microscope objective in which the excitation light passes one side of the objective and the reflected and fluorescence light is collected by the other side as in the Maurice designs. Second, an index-matching liquid (a drop of physiological Ringer's solution) between the cornea and the tip of a nonapplanating objective is used. Third, a computer-controlled piezoelectric micropositioner device scans the microscope objective lens along the z-axis, producing a scan of intensity of reflected light versus distance within the cornea, indicating various cellular layers. This is termed *z*-scanning or axial scanning.

The use of a computer-controlled piezoelectric micropositioner device is the standard technique to control the position of the focal plane of the confocal microscope within a thick specimen. Many microscope objectives are compatible with these micropositioning devices that have high precision along the *z* axis.

A nitrogen laser is the light source, coupled by a fiber optic to the corneal confocal fluorometer microscope. A unique feature of this microscope is the capability to measure simultaneously the scatter profile of scattered light versus depth and the NAD(P)H profile of fluorescence intensity versus corneal depth. As in the Maurice specular microscope, there are two adjustable confocal slits. The depth resolution (axial resolution) is 6 μm with a 100×, NA 1.2 objective, and 18 μm with a 50×, NA 1.0 objective.

This was not the first microscope to use scanning devices to scan the microscope objective with respect to the specimen. Hamilton and Wilson in 1986 published a paper showing a confocal microscope in which the microscope objective was scanned in the plane of the specimen (*x-y* scanning). For very large or heavy specimens, object scanning is not feasible. The solution of Hamilton and Wilson was to keep the advantage of on-axis imaging by using an infinity tube length microscope objective and scanning it in a raster pattern relative to the stationary specimen.

The next step in this developmental sequence was to incorporate many of these instrumental developments into a clinical confocal microscope specifically designed to image the *in vivo* eye, and that is due to Thaer.

8.10 Thaer Real-Time Scanning-Slit Clinical Confocal Microscope

The Koester wide-field specular microscope is capable of photographing wide fields of endothelial cells from the *in vivo* human cornea. However, it has several technical limitations: the custom-built microscope objectives are not interchangeable with commercial ones, the axial optical sectioning capability with the cornea is limited, the applanation microscope objective induces structural folds within the cornea, and the illumination is very bright in the patient's eye.

In order to solve some of these technical difficulties, Thaer in 1994 developed a real-time, scanning-slit *in vivo* confocal microscope (see Fig. 8.8). The optical and mechanical design is based on the previous developments of Maurice and of Svishchev.

The use of a nonapplanating, high-magnification, high-NA, water-immersion microscope objective was adapted from the work of Maurice. An index-matching polymer gel (2.5% hydroxypropyl methylcellulose) is used to optically connect the cornea and the tip of the microscope objective without the need for direct contact. There are two adjustable conjugate (confocal) slits, one for illumination and one for the detection of scattered and reflected light. A two-sided oscillating mirror, similar to that used in the Svishchev confocal design, is used for scanning, descanning, and rescanning for viewing. Also, following the previous designs of Maurice, a split microscope objective is used.

What is unique to this microscope is Thaer's solution to the problem of motion blur. Previous designs attempted to mitigate this problem with the use of a suction

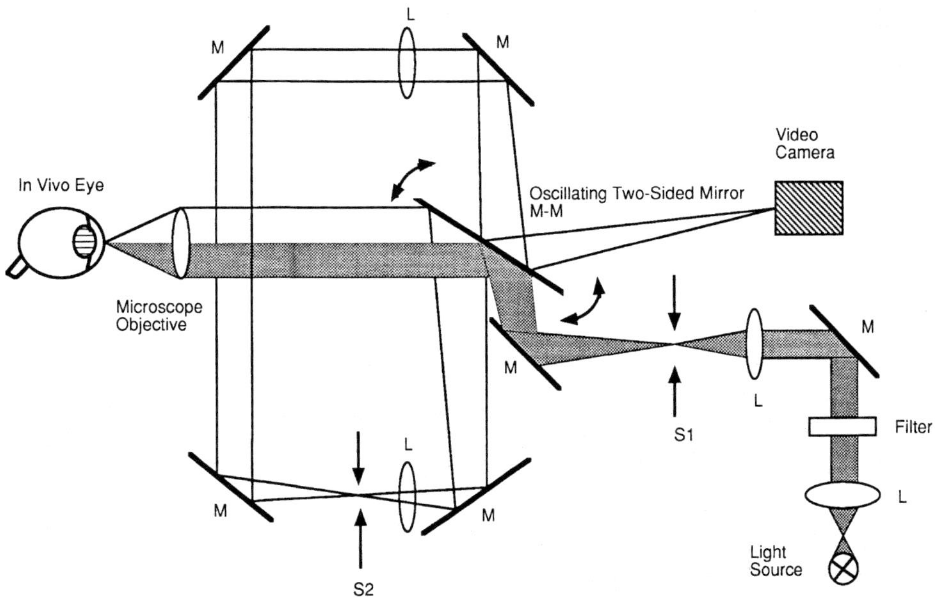

Figure 8.8 Schematic diagram of the clinical, video-rate, scanning-slit, in vivo, confocal microscope with two adjustable conjugate slits: S1 for illumination, and S2 for imaging. An oscillating two-sided mirror M-M scans S1 over the back focal plane of the objective, descans the collected light and directs it to S2, then rescans the beam, which is imaged on the photocathode of a video camera.

ring to stabilize the eye, applanating microscope objectives, or a video camera with frame averaging. Thaer and his collaborators instead used the following technique to eliminate ocular motion blur. Each half of the full video image is formed by a single scan of the slit across the back focal plane of the objective. The slit width is adjusted to 1/30 of the scan amplitude and collects scattered light from the cornea during a period of 0.66 msec. That is, each position of the scanning slit is imaging a region of the cornea for less than a millisecond, and then the adjacent region is imaged for the next 0.66 msec. This technical solution provides high-contrast images through the full thickness of the cornea without any evidence of motion blur. The microscope produces minimal geometrical artifacts; this is demonstrated by imaging the orthogonal lines on the grid from cytometer. In addition, if motion artifacts were significant, the images of nerves would be discontinuous across the full field, and that is not the case. With a 50×, NA 1.0 water-immersion microscope objective, the clinical confocal microscope achieves high-resolution images, high sensitivity, and a lack of motion artifacts. For example, the clinical microscope has the capability to image all the cellular layers within the corneal epithelium.

The clinical confocal microscope has the capacity to produce depth scans (intensity of scattered light as a function of depth within the cornea), or a sequence of images within depth of the cornea (see Fig. 8.9). This function can be achieved by either computer-controlled axial scanning of the objective or the microscope. The

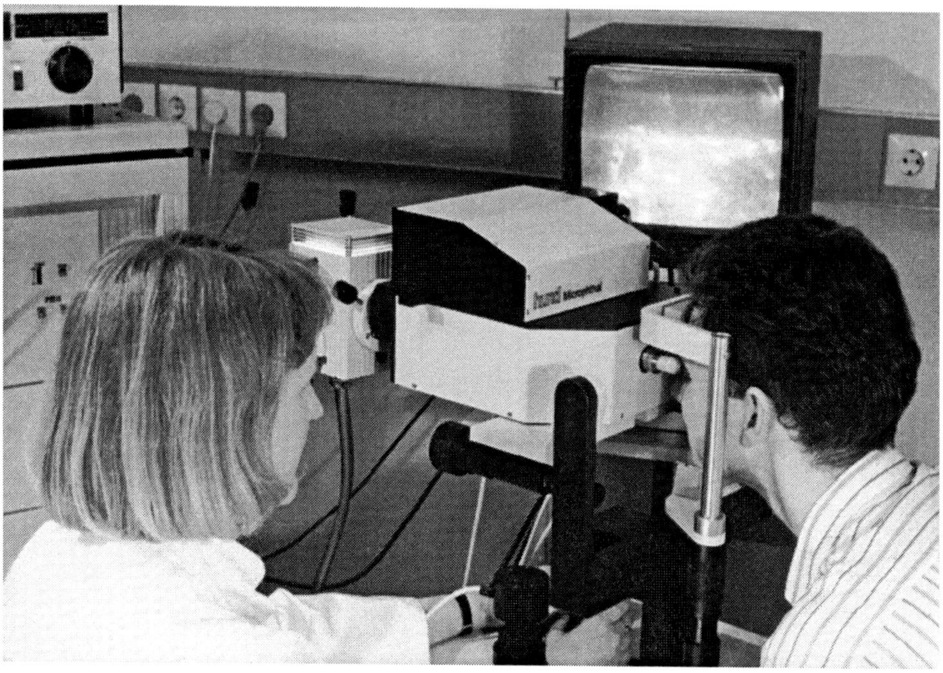

Figure 8.9 The clinical real-time scanning slit confocal microscope.

scanning-slit *in vivo* confocal microscope has been used for many clinical studies and has yielded important results on the normal and diseased cornea.

What are the current limitations of the scanning-slit clinical confocal microscope? The exact axial location of the focal plane within the full thickness of the cornea is not known, and new technical devices are being developed to solve this problem. The clinical confocal microscope uses an index-matching gel between the tip of the microscope objective and the surface of the eye. A totally noncontact imaging system would have many advantages. Finally, the size, weight, and cost of the instrument exceeds the standard slit-lamp instrument.

In this chapter I have highlighted the motivations that resulted in new confocal microscope designs. The clear statement of the problem with the current microscope design was the precursor to the development of a new instrument that provided a partial solution to the previous limitations. I have demonstrated how each inventor was influenced by the advances made by others. The concept of using small slit widths to achieve a thin optical section was combined with the idea of serially building up the composite image as a set of "strips" of a portion of the image plane, and was first used in the confocal instrument of Goldmann. It was incorporated into the confocal microscopes of Svishchev, Maurice, Koester, and Thaer. Similarly, the use of a multisided mirror that performed scanning, descanning, and rescanning was first used in the Svishchev design of a confocal microscope. It was later incorporated into the confocal microscopes of Koester and Thaer. Today, scanning-slit confocal microscopes are finding wide applications in the fields of ophthalmology and dermatology. Their use *in vivo* demonstrates their utility in optical biopsy, and *in vivo* optical microscopy is emerging as an important technique in medicine.

8.11 Summary

- Advances in the designs of scanning-slit confocal microscopes were linked to each other, and were motivated by the necessity to simultaneously provide both narrow optical sections of thick specimens and a wide field of view.
- Scanning-slit confocal microscopes are useful when rapid scanning is important, and when it is necessary to collect more light from the specimen than is possible with scanning systems based on pinholes.
- Advantages of scanning-slit confocal microscopes over point-scanning systems include (1) more light illuminates the sample and enters the detector, (2) the slit width is easily adjustable to compensate for different amounts of light scattering within the sample, and (3) the design and construction of scanning-slit systems is relatively simple.
- Disadvantages of scanning-slit confocal microscopes include (1) the optical sectioning capability is different along the length and the width of the slit, (2) a wide slit will not perform as well as small pinholes with respect to optical sectioning capability, and (3) the full NA of the microscope objective is not utilized in the microscopes described in this section. However, this is not a fundamental limitation. In 1990 Burns et al. invented a scanning slit confocal microscope that uses the full NA of the objective.
- Technology transfer across disparate disciplines is critical for instrument development.
- Svishchev's microscope design, with the technical innovation of scanning the image of a narrow slit over the back focal plane of the microscope objective and descanning the scattered and reflected light with an oscillating two-sided mirror, was incorporated into the wide-field scanning specular microscope by Koester, which was the basis of the bilateral scanning microscope (a reinvention of the Svishchev design), and into the real-time scanning-slit clinical confocal microscope developed by Thaer.
- Svishchev suggested the use of point apertures instead of slits together with continuous-wave lasers and solid state detectors for improved imaging of thick, transparent, light-scattering objects.
- In his patent, Baer suggested two important techniques for three-dimensional microscopy: (1) an objective lens with deliberate longitudinal chromatic aberration that will result in color-coded depth scans formed with a one-dimensional slit scan; and (2) chromatic dispersing prisms in front of the eyepieces that can produce real-time stereo imaging.
- The use of a computer-controlled piezoelectric micropositioner device is the standard technique to control the axial position of the microscope objective.

Chapter 9

The Components of a Confocal Microscope

Many components of confocal microscopes have been described in the previous chapters. In this chapter I will briefly consolidate the key points and provide some additional critical details.

The selection of components should be optimized for the type of specimen to be observed and for the imaging mode, i.e., reflected light or fluorescence. The clinical applications of confocal microscopy to ophthalmology and dermatology are predominantly based on reflected light. However, the biological applications of confocal microscopy are chiefly based on fluorescence techniques. These applications include imaging specific proteins based on green fluorescent protein (GFP) fluorescence, fluorescent indicator dyes that respond to ion concentration and transmembrane voltage differences, fluorescent labeled antibodies to image specific proteins, and fluorescent membrane dyes.

As new fluorescent probes are developed, there may be a need to change the light source to match their absorption bands. Similarly, new dichroic mirrors and filters may be required. New types of detectors are facilitating new techniques; multi-grid photomultipliers in conjunction with fluorescence dispersion devices, and the quantum efficiency and SNR of two-dimensional detectors is improving along with their spectral range. Manufacturers are responding to their users by developing microscope objectives that are optimized for specific applications of confocal microscopy, e.g., neurobiology and developmental biology.

9.1 Light Sources

Light is defined as the visible form of electromagnetic radiation. In the light microscope it is important to make use of these specific characteristics of electromagnetic radiation: wavelength, polarization, and coherence. When choosing a light source one must evaluate broadband or white light sources versus narrowband sources; noise and stability; and the cost of cooling, replacement, and maintenance.

Again, the first question to ask is: What type or types of specimen will be imaged? Is the microscope to be dedicated to imaging integrated circuits? Is it to be used in the clinic to image *in vivo* human skin or the living human eye? What types of fluorescent probes are used with the specimen? What are the absorption bands of the fluorescent probes that will be studied?

White light sources are necessary for real-color direct-view Nipkow disk-based confocal microscopes. Broadband light sources are obtained from arc lamps together with sets of filters. Nonlaser light sources have a lower degree of coherence as compared with lasers. It is desirable to have a low-coherence source of light for reflection confocal microscopy, to reduce interference and speckle in the specimen; high-coherence lasers may cause fringes to develop in the image if there are multiple reflections. When necessary, there are methods available to break the coherence of a laser light source, e.g., vibrating a fiber optic that transmits the laser light.

A variety of light sources can be used with various types of confocal microscopes. Several Nipkow disk-based confocal microscopes use arc lamps as the light source. A halogen lamp is the light source in the scanning slit clinical confocal microscope. Light-emitting diodes are a good alternative light source for Nipkow disk and scanning-slit confocal microscopes; they are stable, efficient, and have long lifetimes.

Light sources for confocal microscopes can be divided into spatially coherent or spatially incoherent groups. LSCMs are spatially coherent. In contrast, many of the clinical confocal microscopes used in the clinic, for example, the scanning-slit confocal microscope used in the ophthalmology clinic uses halogen lamps as a spatially incoherent light source. With spatially incoherent illumination, the phase relations between fields at nearby points are statistically random.

Spatially coherent light sources have the important property that the phase difference between any two points is constant with time. Examples of spatially coherent light sources are lasers and arc lamps with a small aperture that acts as a spatial filter. There is another important term: **temporal coherence**. A laser with a single frequency (actually a very narrow range of frequencies) would have a high temporal coherence. That term implies that there exists a definite phase relationship between the fields at a given point after a time delay of T. Usually practical lasers show this definite phase relationship for a fixed time, called the coherence time.

Lasers usually have a single wavelength or they can be tuned to output a few discrete wavelengths. Another possibility is to use two or three different lasers to provide a wider selection of wavelengths. Most modern LSCMs use one or more lasers as the light source. The output from a laser is extremely bright, monochromatic (a single color, with a very narrow range of frequencies), coherent, and highly collimated. The output is typically also linearly polarized; this can be exploited in differential interference contrast microscopy, polarized light microscopy, and studies of fluorescence polarization anisotropy. A high-NA microscope objective can focus the laser beam to a diffraction-limited point or volume of light that is extremely intense.

The laser beam may be expanded in diameter before it is coupled to the confocal microscope. Typically, the diameter of the laser beam is adjusted to overfill the back aperture of the microscope objective. When the laser operates in the TEM_{00} mode, it forms a Gaussian beam; in that case an illumination pinhole aperture is not needed, because the light appears to originate from a point source at infinity. The laser beam may be expanded. With a high-NA microscope objective, a diffraction-limited spot is formed at the focus.

For confocal microscopes, e.g., Nipkow disk or reflected light confocal microscopes, in which the coherence of the laser light is not desirable since it can cause speckle and other problems in the image, it is possible to degrade the coherence. One method is to pass the laser light through a piece of optical fiber that is forced into vibration. The vibration caused rapid changes in the optical path length that makes the beam temporally incoherent. Another method is to place a rotating glass wedge in the beam path.

A laser can operate in a variety of modes; the transverse electromagnetic mode that is called TEM_{00} has some unique characteristics. The emitted light has a quasi-planar wavefront with a Gaussian intensity profile perpendicular to the direction of the beam propagation. The TEM_{00} mode is also important because it has minimum beam divergence and can be focused into a diffraction-limited spot. Some lasers, such as helium neon or argon ion lasers, have TEM_{00} outputs that are cylindrically symmetrical.

A helium neon laser produces a **Gaussian beam** of laser light, which is a beam of light in which the electric field distribution is radially symmetrical and can be mathematically described by a Gaussian function of two parameters: *r*, the radius of the circular beam, and *w*, the **beam waist**, i.e., the smallest diameter of the beam. A Gaussian beam of light in cross section (perpendicular to the direction of light propagation) would appear as a bright spot of light without any nodes or regions of zero intensity. The Gaussian beam is a result of the type of mirrors in the laser cavity within the HeNe that produced the laser beam.

A Gaussian beam has several very interesting and unique properties. First, the Fourier transform of a Gaussian function is also a Gaussian function. Second, as a Gaussian beam passes through an optical system consisting of many lenses or a single lens, the intensity distribution remains Gaussian at every point. With a weakly focused Gaussian beam of light, the beam waist does not occur at the focal length of the lens!

There is a potential danger of a highly focused laser beam; the power densities can be extremely high and can result in eye and skin damage as well as destruction of optical components such as lenses, optical fibers, and optical coatings. Of course, the same high power densities can damage the specimen. A power density of 30 million watts per square centimeter is easily obtained from a 10 mW laser beam. All that is required is to expand the laser beam and then use a lens to focus the beam down to a diffraction-limited spot of about 0.2 μm.

HeNe lasers are not expensive, have a long lifetime, and are stable sources. They output a few milliwatts of power and are available in red (632.8 nm), green (543.5 nm), yellow (594.1 nm), and orange (611.9 nm). HeNe lasers have excellent beam quality; moreover, they are purely Gaussian. HeNe lasers are compact, convection-cooled, and vibration-free. The combination of red, green, and blue lasers can be used to produce true-color confocal microscopy.

Typically, the color of a light source is given in terms of its wavelength. Alternatively, we can relate the color of a light source to the frequency of the light. The frequency of the light is directly proportional to the energy of its photons. The visible spectrum of light has frequencies in the range from 0.4 PHz to 0.7 PHz (petahertz, equal to 10^{15} Hz).

The helium-cadmium laser is useful for the production of several lines in the blue (443 nm) and ultraviolet region (325 nm and 354 nm). The emission at 442 nm is useful for the excitation of flavins and other fluorescent molecules. It is suggested that a laser stabilization device based on an acousto-optic device be employed to improve laser stability.

Argon-ion lasers are commonly used to excite fluorescent probes with confocal microscopes. Air-cooled argon-ion lasers usually contain a cooling fan that is a source of vibration that can easily be transmitted to the confocal microscope. It is suggested that the cooling fan be removed and situated at a distance from the laser. A lamellar tube, not the corrugated laundry duct flexible tubing that generates turbulence and vibration, should be used to connect the fan and the laser. Argon-ion lasers have several other disadvantages: high cost, limited laser-tube lifetime, and heat production.

Argon-ion lasers can output several laser lines in the range of 50–500 mW. The wavelengths include violet (454 nm), blue (488 nm), and several others. The air-cooled krypton-ion laser can output red (647 nm) and yellow laser light (568 nm).

Another useful light source is the mixed-gas argon-krypton laser. This laser can produce several laser lines (the wavelength of both lasers combined) across a wide range of the spectrum and offers a cost reduction compared with the purchase of two individual lasers.

Alternatively, diode-pumped solid state lasers can output 50 mW of power at 532 nm, which is in the green region of the spectrum. A diode-pumped neodymium-yttrium aluminum garnet (Nd:YAG) laser can generate milliwatts of power at 1064 nm. Frequency doubling can produce a laser output at 532 nm, and frequency tripling can produce an output at 355 nm.

Semiconductor or diode lasers are another source of light for confocal microscopes. They are formed as a junction diode that is made from type-n and type-p semiconductor crystals separated by an undoped semiconductor. Semiconductor lasers are made with several different compounds, e.g. GaAs, and operate on the principle of producing laser radiation as a result of electron-hole recombination. Some of the diode lasers produce TEM_{00} output beams that are asymmetric and often astigmatic.

Other diode lasers are available with outputs of light at 406 nm and 440 nm. An alternative light source for the argon-ion laser is an InGaN violet laser with an output at 406 nm. With this laser source it is possible to image cells stained with fluorescent dyes that normally require excitation ultraviolet argon laser lines, such as green fluorescent protein.

While one may think that increasing the power of the illumination will result in more intense images with an increased SNR, two important processes must be considered: **fluorescence saturation** and **photobleaching**. When the rate of absorption of a fluorescent molecule exceeds the rate at which the energy from the excited state can be released by either radiative processes, such as fluorescence, or nonradiative processes, such as singlet-triplet transfer or heat production, then we have the phenomenon of light saturation. Further increases in the intensity of illumination will not increase the intensity of fluorescence.

Photobleaching occurs when the excited state of the fluorescent molecule reacts with oxygen to produce a photochemical reaction in which the fluorescent molecule is transformed into a nonfluorescent molecule. Over a period of time with constant illumination of a volume in the specimen, it will be observed that the fluorescence intensity is reduced. Therefore, a high intensity of illumination may be damaging to the fluorescent molecules. Lowering the illumination intensity will only lower the rate of photo-destruction of the fluorescent molecule; it will not completely eliminate the process.

Cell biologists now have a wide range of fluorescent probes that span the wavelength spectrum. Often it is necessary to use several different lasers to obtain the appropriate wavelengths in order to work with a variety of fluorescent probes. One interesting approach is to use a Ti:sapphire laser to pump a photonic crystal fiber that outputs a visible continuum of light. This continuum of light can then be selectively filtered to provide any of the required peak excitation wavelengths (McConnell, 2004).

9.2 Scanning Systems

Previous chapters have described the Nipkow disk-based confocal microscopes, which use multiple beams of light to scan the focal place. Also, Chapter 8 described the development of scanning-slit confocal microscopes. A slit can be thought of as a linear array of adjacent pinholes. Therefore, the scanning-slit confocal microscopes can also be placed in the category of multiple-beam scanning systems. This section describes and compares two types of single-beam scanning. First, in specimen or stage scanning the microscope objective is used on-axis and the stationary illumination is confined to the center of the lens. What is scanned is the stage or the specimen; this is achieved by laterally moving the specimen in the focal plane of the objective relative to a stationary optical path. Second, in beam scanning, an angular motion of the illuminating beam fills the back focal plane of the microscope objective. That movement causes the focused light beam to be displaced laterally in the focal plane relative to the stationary specimen.

Specimen scanning has several distinct and important advantages. First, the optical system is simple, and it must only produce an axial diffraction-limited spot of light. Since we only use the axial region of the microscope objective, many off-axis optical aberrations are eliminated or minimized; only the spherical aberration of the lens is to be considered. Second, the microscope objective is only used at one part of its field, so field curvature is not important. Third, the resolution and the contrast are identical across every region of the specimen; the illumination is constant with scanning. Fourth, the resolution and the contrast are independent from the magnification; there is **space invariant imaging**. Fifth, in principle, large specimens can be imaged, e.g., a very large wafer containing integrated circuits.

Specimen or stage scanning does have its limitations. A possible disadvantage of this system is the relatively slow speed of image acquisition: on the order of a few seconds, depending on the size of the object to be imaged. There are imaging situations in which rapid image acquisition, for example, in calcium imaging in excitable tissues, is a necessity, which would not work with stage scanning. If rapid image acquisition is a requirement, then the mass of large specimens may be a constraint.

Yet another method is to scan the microscope objective laterally over the specimen in a plane parallel to the focal plane within the specimen. With **lateral objective scanning**, the illumination overfills the microscope objective. This technique also has a slow speed of image acquisition.

Many of the commercial confocal microscopes use a **beam-scanning** system. Beam scanning is not space invariant. Various scanning systems are used to scan the light beam over the back focal plane of the microscope objective. Either a diffraction-limited spot or a slit of light can be scanned over the back focal plane of the objective. Several methods can be used to achieve beam scanning: vibrating galvanometer-type mirrors, rotating polygon mirrors, and acousto-optic beam deflectors.

An alternative arrangement to the two galvanometer mirror scanners for deflecting the laser beam in the x and y directions is to use a single mirror that tilts or oscillates around two orthogonal axes. This is achieved by mounting a single mirror on a rapid galvanometer, which is mounted on a second orthogonal scan system, with its center of rotation in the center of the mirror.

Very high frame rates can be achieved by combining a scanning mirror on one axis (relatively slow) with a rotating polygon mirror on the orthogonal axis (very fast). For the case of beam scanning, the magnification is now coupled to the resolution; that is, the imaging is not space invariant. Several different microscope objectives are normally required to cover a range of magnifications. A beam-scanning confocal microscope can easily be constructed around a conventional microscope.

An ideal laser beam scanning system will bring the laser beam into the microscope with the constraint that at each scan position, the laser beam is focused to a point on the specimen. At every point of the scan, the back focal plane of the microscope objective should be filled with the laser beam. The only way to achieve this condition is to rotate the laser beam about the **stationary pivot point** imaged in the back pupil of the microscope objective. As the laser beam is scanned over the specimen, the part of the laser beam on the rear pupil of the lens should remain stationary and rotate around a pivot point.

The Bio-Rad (now part of Zeiss) laser beam scanning confocal microscope has the following solution to conform to beam scanning with a stationary pivot point. The laser beam is directed to the first oscillating galvanometer mirror. The laser beam from the first mirror is imaged onto the second oscillating galvanometer mirror with two concave reflecting surfaces. The beam scans over the two concave reflecting surfaces, but is always directed to the same point on the second mirror. This optical arrangement results in a scan that is a pure rotation about a point on the second mirror.

Typically, in a beam-scanning system an afocal optical relay is used to image the first galvanometer mirror onto the second galvanometer mirror. A second afocal optical relay is then used to image the second galvanometer mirror into the back focal plane or objective aperture of the microscope objective.

A typical LSCM displays an image of 2048 × 2048 pixels. The frame rate will of course depend on the signal intensity and the noise levels. The frame rates are slowest for the 2048 × 2048 images. In order to increase the frame rate and decrease the image acquisition time, there are two options: either reduce the number of pixels in the frame or acquire a line scan over the specimen. A line scan is a linear array of pixels.

An alternative approach to achieve rapid beam scanning is to use an acousto-optic deflector or modulator, to deflect the laser beam. This device is sometimes called an acousto-optical tunable filter (AOTF), which was described in Sec. 7.8. When used for confocal reflection imaging, the wavelength of the illumination light and the scattered and reflected light are identical. Therefore, the light from the specimen can be descanned back through the acousto-optic deflector or modulator. For fluorescence confocal microscopy, the acousto-optic deflector cannot be used since the longer-wavelength fluorescence emission cannot be descanned back through the device, as it is wavelength specific.

All these methods of single-beam scanning confocal microscopy are limited by the following factors: image acquisition time, SNR, and the spatial resolution of the scan. Rapid or video-rate image acquisition may at first seem advantageous in all cases; however, the resulting images may be degraded and limited by a low SNR.

9.3 Dichroic Mirrors and Filters

Reflected-light confocal microscopy dominates clinical applications in dermatology (imaging *in vivo* skin) and ophthalmology (imaging *in vivo* cornea). Nevertheless, it is the fluorescence mode of confocal microscopy with its spectacular sensitivity and specificity that is widely used in biological and medical applications.

As described in Chapter 1, early fluorescence microscopes used transmitted light for fluorescence. These diascopic designs had the light source, excitation filter, condenser, specimen, objective, emission filter, and the detector in a line. Since the intensity of the excitation light is thousands of times greater than the fluorescence intensity, problems occurred with noise because some of the excitation light contributed to the "fluorescence" image. This problem was mitigated but not solved with the development of dark-field condensers that illuminated the specimen with oblique light.

Modern confocal microscopes that operate in the fluorescence mode use an episcopic or epi-fluorescence design in which the excitation light source is perpendicular to the path of the fluorescence. The development of dichroic mirrors or beamsplitters, which separate the excitation light from the emission light (longer wavelength), resulted in enhanced utility and consequently universal acceptance by the biological community. When one compares a nonconfocal fluorescence with a confocal fluorescence image of a thick specimen, the different is striking. While the development of fluorescent probes also contributed to the growth of biological applications of fluorescence confocal microscopy, the development of the dichroic beamsplitter or mirror was a major component to this growth.

When the confocal microscope is used in the fluorescent mode with vertical illumination, it is necessary to separate the exciting light from the fluorescence light before the light reaches the point detector. This function is typically performed with a **dichroic mirror** and filters. The dichroic mirror reflects all wavelengths shorter than a fixed wavelength and transmits all wavelengths longer than the threshold. The microscope can have several different channels available (several wavelengths of excitation light) and a corresponding number of detectors. Dichroic mirrors can reflect more than 95% of the excitation light and transmit more than 90% of the fluorescence emission light. Typically, the dichroic mirror is the center element of three filters: the excitation filter, the dichroic mirror, and the barrier filter. All three filters are usually combined into a single fluorescence filter cube or block.

The selection of the correct dichroic mirror will depend on the absorption band of the fluorescent molecule, the spectral properties of the light source, and the spectral sensitivity of the detector. As shown in Fig. 2.4, the dichroic mirror is inserted at a 45-deg. angle to the optic axis of the microscope.

An alternative is to use one computer-programmable AOTF as the excitation filter, and another in place of the traditional dichroic mirror. The advantages of using the AOTF include speed of changing the transmission band and also rapid intensity control of the laser beam. The downside is that the fluorescence cannot be descanned by device. An AOTF could be used to replace dichroic mirrors for wavelength selection that is programmable over a wide range of wavelengths. When used with multiple laser lines as the light source, two AOTFs provide rapid control of the excitation and the emission wavelengths. These solid state devices can modulate the intensity, deflection, and transmission of multiline light sources and may offer advantages over beam scanning based on oscillating mirrors and wavelength selection based on sets of dichroic mirrors.

9.4 Pinholes

Pinholes are apertures placed in front of the light detectors, converting an area detector into a point detector. In a confocal microscope, both the point source of light and the pinhole in front of the detector should be cofocused. This is the **confocal principle**. Another way to state this is as follows: in a confocal microscope the confocal aperture, which is called a pinhole, is located in a plane conjugate to the intermediate image plane and, thus, to the object plane of the microscope.

As the pinhole size is increased, which means that the detector size is also increased, a confocal microscope becomes more and more like a conventional microscope. With very large pinholes in front of the light detector, the axial resolution of the confocal microscope is lost and light from out-of-focus planes will contribute to the image. A fluorescence (incoherent light) confocal microscope has a resolution that is about 1.4 times greater than that of a conventional light microscope. This optimal resolution is achieved if the confocal pinhole, located in front of the detector, is smaller than the Airy disk formed from a diffraction-limited point of fluorescence in the specimen.

As the pinhole in front of the detector of a LSCM is made smaller, there are two effects: first, the resolution is increased; and second, the signal from the detector is reduced. If the pinhole size is set to be smaller (50–75%) than the first minimum of the Airy disk (the image of a point source of incoherent light), then a good compromise is achieved between signal strength and the degree of background rejection. Computer modeling suggests the following: first, the lateral (transverse) resolution is more sensitive than axial resolution to pinhole size; second, resolution differs according to whether the object is a point, line, or plane; and third, aberrations in the microscope objective, mainly spherical aberrations, will skew the shape of the Airy disk so that it becomes asymmetrical about its center.

The microscope objective images as an Airy disk a diffraction-limited spot of incoherent light (i.e., a subresolution fluorescent bead) in the image plane of the light microscope. The size of the Airy disk will be proportional to the wavelength of the light used in the illumination, and inversely proportional to the NA of the microscope objective. If the pinhole is adjusted to a size just slightly smaller than the first minimum of the Airy disk, there will be optimal resolution and background rejection. For some very weakly fluorescent specimens it may be necessary to enlarge the pinhole, i.e., to make it larger than the optimal calculated size, in order to increase the signal and therefore the SNR in the image.

In practice, several factors will degrade both the signal and resolution compared to the ideal instrument. It is critical that the center of the pinhole be aligned with the Airy disk. For a low-magnification lens with a small Airy disk, this is especially important. Another very important factor is that the illumination beam must fill the back focal plane or aperture of the microscope objective. The size of the back focal plane differs for various microscope objectives with different magnifications. It may be necessary to expand the laser beam to fill the back focal plane of a low-magnification microscope objective. Depending on whether the confocal microscope is based on a Nipkow disk, a laser-scanning system, or a scanning-slit system, the type and arrangement of the pinholes will differ.

We have discussed the Nipkow disk-based microscope. In this confocal microscope the pinhole size is fixed for a given disk. The size of the pinholes is designed for a given resolution and their spacings are designed to eliminate cross-torque between adjacent pinholes. If the pinholes are too closely spaced, there could also be interference between their images on the specimen. With a narrowband light source, there could be interference and speckle effects. Speckle results from the interference of light scattered by adjacent spots within the illumination field. Typically, the pinhole dimensions are in the range of 20 to 80 μm.

The very low light throughput in this type of confocal microscope is because of the very small ratio of area of the pinholes to the area of the disk. Typically, the area of the pinholes is less than 1% of the illuminated area of the disk. A similar situation exists for a confocal microscope based on a one-sided Nipkow disk. As previously described, these severe limitations can be overcome with microlens arrays. The light throughput can be as large as 40%.

We now discuss the pinholes in the LSCMs. Again, two optical designs exist. In the original BioRad design, the optical path of the light from the specimen is about 2.5 m, and therefore the Airy disk is greatly magnified to the diameter of several millimeters. In that case, instead of a pinhole it is possible to use an adjustable iris in front of each detector. Each channel would have its own adjustable iris in front of its light detector. This offers several advantages: the adjustable iris is continuously adjustable, not sensitive to dust and dirt, and cheap to make. Other manufacturers of LSCMs use fixed pinholes on a slider. Although the pinhole size can be changed, the choices are a set of discrete diameters. The very small pinholes are susceptible to vibration and dirt.

Finally, confocal microscopes based on scanning slits offer their own characteristics. An adjustable slit is made to a fixed length but with a continuously variable width. A confocal microscope based on slits is only truly confocal in the direction perpendicular to the length of the slits; hence, the resolution in the plane of the specimen differs depending on the orientation of the slits. The resolution is best in the direction perpendicular to the length of the slits, and worse in the orthogonal direction. At the same time, variable slit width has a great advantage in a confocal microscope. For weak specimen signals it is possible to open up the slit width and trade off resolution for increased signal. This is very useful for clinical imaging of the living human eye and *in vivo* human skin.

9.5 Detectors

The eye is the detector with direct-view, Nipkow disk-based confocal microscopes. For reflected-light microscopes, the image quality is excellent, the image acquisition time is video rate or faster, and the images are observed in real color. There is no permanent record of the image. Today, confocal microscopes display images digitally on a monitor, which means that they can be digitally processed in a computer, stored in digital form, and shared with anyone who has access to the Internet. Remote microscopy is now a reality.

What are the general characteristics of detectors that are critical for their use in a confocal microscope? The type of confocal microscope will determine whether a point detector such as a photomultiplier tube or a two-dimensional detector is required. The eye can be replaced with a sensitive, cooled, low-noise, high-dynamic-range, real-color CCD camera to give a digital image with direct-view, Nipkow disk-based confocal microscopes.

The signal strength at the detector will set the constraints for the detector selection. When the signal is strong, many confocal microscopes use analog detection techniques. However, when the signal is extremely low, techniques for single photon counting are required (Becker, 2005). The important rule in detection is that every photon counts; it is important to maximize the collection and detection of all photons from the specimen. In the reflection mode of confocal microscopes, the image is formed from the reflected and the scattered light. The use of various wavelengths could alter the penetration depth of thick specimens and also the contrast of the images. In the fluorescence mode, it is important to use barrier filters or dichroic mirrors to isolate the fluorescence light of the specimen from the excitation light. The use of coatings on the optical elements and the careful design of the optical system to reduce stray light are critical. Since each optical element contributes to the flare and stray light, it is an important design consideration to minimize the number of optical elements in the microscope.

Confocal microscopes can be equipped with several types of detectors: photomultiplier tubes (PMTs), avalanche photodiodes (APDs), and charge-coupled-device (CCD) cameras. A clear but dated review of solid state detectors and cameras is Chapter 7 of *Video Microscopy, 2nd ed.* (Inoué and Spring, 1997).

The most common detector used in confocal microscopy is the PMT since it is relatively inexpensive, sensitive, and stable. Photomultipliers have high sensitivity, ultrafast response, high bandwidth, and a gain that is approximately 10^6.

PMTs are commonly used for low-level light detection. A typical photomultiplier tube contains a photocathode that converts incident photons to electrons, a number of amplifying stages called dynodes, and an anode. Photomultiplier tubes have a very high gain or amplification. In the absence of incident light, the dark count represents unwanted background signal. By cooling the photomultiplier tube, it is possible to reduce the dark counts.

PMTs have a quantum efficiency—i.e, the detection fraction of incident photons—that depends greatly on the wavelength of the light, varying from 1% to 20%. The photocathode of a PMT should be selected to match the wavelengths of the fluorescence for a given fluorescent probe. The types of materials in the photocathode determine the radiant sensitivity. It is important to select a photomultiplier with a photocathode that has the appropriate light sensitivity for the wavelengths to be detected. The bialkali and multialkali photocathodes of conventional photomultipliers are usually in the range of 20% in the wavelength range of 400–500 nm.

It is possible to acquire the spectrum at each pixel scan position. The fluorescence from the specimen is first separated from the illumination light with a dichroic mirror, then directed to a dispersive element such as a prism or a grating. The dispersed light is simultaneously detected by a multi-anode PMT. These devices have eight or more anodes that act as independent photodetectors. The digitized output from each individual anode element, which corresponds to a specific spectral band, is acquired for the eight spectral regions at each pixel. Therefore, the spectrum for each pixel of the image is recorded. This spectral data are very useful for the application of fluorescence techniques in cell biology, e.g., FRET.

An APD, which is a semiconductor device, is an alternative to a PMT. APDs have a very large gain in the conversion of photons to electrons. They have a quantum efficiency of about 80% in the wavelength of 550–750 nm, and the quantum efficiency falls off at shorter wavelengths. The important limitation is that APDs do not have the dynamic range of photomultipliers, and therefore they can easily become saturated with bright light. Nevertheless, for low-light situations, APDs may be the best choice. For detection of light in the near infrared, the use of a single-photon avalanche photodiode (SPAD) with a quantum efficiency of 80% at 700 nm may offer advantages over PMTs.

Recently, significant technical advances have occurred in two-dimensional charge-coupled-device (CCD) cameras. These highly sensitive, low-noise cameras are useful for low-light-level fluorescence imaging. For example, the Cascade II: 512 camera from Photometrics operates at –80° C, which minimizes the dark current. The CCD chip uses on-chip multiplication gain (1–1000) which yields very high sensitivity and low noise. These detectors use the on-chip multiplication gain to multiply the photon-generated charges above the readout noise, even at and above video frame rates.

The back-illuminated image sensor has a quantum efficiency of over 90% at the peak of 500–700 nm. It offers 16-bit digitization, which results in a high dynamic range. A high dynamic range allows the detection of both dim and bright signals in the image.

In a standard-use fluorescent confocal microscope that contains three channels, the best combination of PMTs may be one tube with an S20 photocathode and two tubes with bialkali photocathodes; that would cover the wavelength range useful for a majority of the commercial fluorescent probes used in confocal microscopy.

For extremely low level fluorescence of a specimen, the photomultiplier signal can be processed with electronics that operate in the photon-counting mode (Becker, 2005). When the specimen is stable and the image can be built up over several seconds, the photon-counting mode is a good choice. Photon counting only counts those photons that are above a given light-intensity threshold; therefore, the image is built up from a dark background. The limitation of the technique is that it is not suitable for rapid image acquisition.

An important consideration is the role of noise in the detector and its associated amplifier. Remember, a light detector detects intensity. Several components contribute to the total noise in a light detector. First is photon shot (Poisson) noise, which is a direct consequence of the quantized nature of light. Note that discrete photons comprise the light beam. Photons in the beam impinge upon the light detector at randomly distributed time intervals. Photon shot noise results from the Poisson statistics of counting photons. If the light beam has N photons per unit of time, then the number of photons measured will vary around the average value with an amplitude equal to the square root of N.

Second, at temperatures greater than absolute zero, the detector will emit electrons even in the absence of photons. That spontaneous noise is called **dark noise**, which is caused by the thermal generation of electrons in the detector. Cooling the detector can reduce the magnitude of the dark noise. Third, the photons that enter the photodetector cause electrons to be emitted. The electrons are passed across a resistor to become a voltage measured in the amplifier. Since resistance fluctuates as a function of temperature, there is another source of noise, called Johnson noise. Fourth, laser noise occurs, caused by the random fluctuations in the filling of excited states in the laser medium. Laser noise is proportional to the signal amplitude.

All these sources of detector noise contribute to the total noise in a light detector: the total noise is the square root of the sum of each source of noise squared. The signal to noise ratio is the number that will determine the quality of the image derived from the confocal microscope. Several sources of noise exist, including the quantum nature of the light. In general, as the number of detected photons (N) increases, the signal to noise ratio will be enhanced by the square-root of (N).

A typical LSCM will have four independent channels, which means each probe can be imaged simultaneously. With 12-bit analog-to-digital converters, there are 4096 levels of brightness in the digital image.

9.6 Microscope Objectives

The microscope objective is a key component in a confocal microscope. While it has been designed and manufactured to extreme tolerances, it is subject to dirt and mechanical damage that can severely degrade its optical qualities.

Therefore, the selection, care, cleaning, and use of a microscope objective are critical. It is suggested that the user of a confocal microscope study the websites of the major manufacturers such as Zeiss, Nikon, Olympus, and Leica for the latest information on the care of microscope objectives. For example, the use of tissue paper to clean a microscope objective will result in permanent damage to the optical surface!

The selection of an appropriate microscope objective will depend on the magnification required, the use of a cover glass of the correct thickness for the particular microscope objective, the NA of the microscope objective, and the free working distance required. Other factors to be considered are the various types of aberrations. Often a large refractive index mismatch exists between the specimen and the optical system, which consists of a layer of index-matching oil, a cover glass, and the microscope objective. This index mismatch can result in large aberrations of the optical system and a loss of image fidelity between the specimen and the resulting image.

Typically, high-NA oil-immersion microscope objectives are corrected for viewing specimens located close to the coverslip. When these objectives are used to image thick specimens such as tissue, the refractive index mismatch creates large uncorrected spherical aberrations. A solution to this problem is to use high-NA water-immersion microscope objectives without a coverslip.

In recent years, several manufacturers have produced high-quality water-immersion microscope objectives with long working distances, high NA, and high magnification. For the optical observation of thick living specimens, an optimal solution may be the use of long-working-distance water-immersion microscope objectives without the use of a cover glass. Many modern confocal microscopes use infinity-corrected microscope objectives. An important advantage of infinity-corrected optics is that the focal plane can be changed by moving the position of the objective rather than having to displace the microscope stage.

Studies of live cells are performed with a light microscope in a controlled environment at 37° C. Leica manufactures a 63×/1.3 microscope objective that is corrected for 37° C and balances for mismatch in refractive index, cover glass thickness, and temperature. Olympus manufactures a microscope objective specifically designed for ultraviolet confocal microscopy. It is corrected for the ultraviolet from 350–650 nm. One version is an infinity corrected objective, water-immersion, 40×, that focuses ultraviolet light and blue light to the same point as visible light.

When confocal microscopy is used with cells in culture, it is often necessary to use micropipettes and electrodes. This type of study requires that the microscope objective have a long working distance. An example of these newly developed low-magnification, high-NA, long-working-distance water-immersion microscope objectives is manufactured by Olympus. This infinity-corrected microscope objective is a 20× water-immersion objective with an NA of 0.95 and a working distance of 2.0 mm.

The selection of a microscope objective usually involves a trade-off of several variables. The user must decide what parameters are critical to the current imaging problem and make the appropriate selection. There are several important characteristics of a microscope objective for use with a confocal microscope: NA, free working distance, correction for spherical aberration, and correction for chromatic aberration.

For example, if it is critical to have the brightest image, then a good choice would be a microscope objective with a low M and a high NA, such as 40×/1.3. The size of the back pupil of the microscope objective sets the limitation when used in the epi-fluorescence mode. Another factor is the loss of light due to reflection within the microscope objective. Microscope objectives with magnifications of 60× and 63× and NAs of 1.2 to 1.4 have larger back pupils than the 100× objectives of the same NA, and yield brighter images.

Another factor is the number of lenses in the objective; the lower the number of lenses, the brighter the image. Oil-immersion microscope objectives use oil to achieve the high NA value. It is prudent to check the fluorescence of the sample of immersion oil before using it. Finally, check the spectral response of the microscope objective. Although today it is possible to obtain microscope objectives with a wide range of features, there is still a range of optical properties, even with the same catalog part number. If possible, try several of the same kind of microscope objectives and select the one that maximizes the optical properties that are of highest interest for the particular specimen.

Furthermore, even when the microscope objective is correctly selected and everything is perfectly aligned in the optical system, an incorrect refractive index mismatch between the microscope objective, the oil objective, and the specimen could result in a strong spherical aberration with a resulting loss of resolution and contrast.

Microscope objectives designed for use with confocal microscopes should have a long working distance to cover thick specimens, a high NA for high optical resolution, a high brightness of the image, a planar image field, and very low chromatic longitudinal aberrations. The last two characteristics are important for correct three-dimensional reconstructions.

An objective with low chromatic longitudinal aberrations is only an important consideration in a scanning laser confocal microscope when there are several lasers (or one laser with multiwavelength output) and several fluorophores. Most modern fluorescent confocal microscopes have several fluorescent channels; but if the user is only using a single wavelength and a single fluorophore, then chromatic aberrations are less important.

Many researchers find that long-working-distance, water-immersion microscope objectives are extremely useful. The objective tip is inert, corrosion-free, chemically resistant, and has very low thermal conductivity. These microscope objectives have found widespread use for *in vivo* confocal microscopy of the human eye and skin, and live cell and tissue imaging.

The invention and development of the confocal microscope offers a partial solution to the original problems that have plagued light microscopy. The severe problems of resolution and contrast have been partially solved. The confocal microscope is a direct solution to the problem of out-of-focus haze in the conventional fluorescence microscope. The confocal microscope is readily available in many laboratories in a variety of instruments, from the laboratory bench top to the clinic, where it is used in dermatology and ophthalmology.

The development of the confocal microscope left unsolved the following problems: how to perform fluorescence microscopy with fluorescent probes or autofluorescence excited in the ultraviolet; how to increase depth penetration into thick, highly scattering tissues; and finally, how to minimize the deleterious and sometimes lethal effects of long-term microscopy on living cells and tissues and organs. Also, the depth of penetration is limited by the absorption and scattering of the illumination light. Highly adsorbing and scattering regions within the specimen will severely reduce the intensity of the illumination at more distal regions of the specimen. For highly transparent specimens such as the *in vivo* human cornea, the depth of penetration can exceed 550 μm.

In order to partially solve these important problems, we turn to a new form of nonlinear microscopy: multiphoton excitation microscopy. Part III develops the background of nonlinear spectroscopy, traces antecedents to the invention of multiphoton microscopy, and discusses the new microscopes, the theory of multiphoton microscopy, and its limitations.

9.7 Summary

- A wide variety of laser sources are available: blue argon (488 nm), multiline argon (351 nm, 457 nm, 488 nm, 514 nm), green helium neon (543 nm), orange helium neon (594 nm), red helium neon (633 nm), yellow krypton (568 nm), helium cadmium (442 nm), and diode lasers (406 nm, 440 nm).

- Noise is a fundamental problem in detectors. Very weak signals may be improved by increasing the light intensity of the illumination or the integration time of detection.
- A fluorescence (incoherent light) confocal microscope has a resolution that is about 1.4 times greater than that of a conventional light microscope. This optimal resolution is achieved if the confocal pinhole located in front of the detector is smaller than the Airy disk formed from a diffraction-limited point of fluorescence in the specimen.
- As the pinhole in front of the detector of a LSCM is made smaller, there are two effects: first, the resolution is increased, and second, the signal from the detector is reduced. If the pinhole size is set to be slightly smaller than the first minimum of the Airy disk (the image of a point source of incoherent light), then a good compromise is achieved between the signal strength and the degree of background rejection.
- Fluorescence saturation and photobleaching of the fluorescence are limits in fluorescent confocal microscopy.
- Solid state, programmable acousto-optical modulators could replace traditional dichroic mirrors.
- Although various scanning systems are available, some with very high frame rates, the limitation is the signal to noise ratio of the detected signal.
- Specimen or stage scanning has some distinct advantages. The optical system is simple and must only produce an axial diffraction-limited spot of light. Since we only use the axial region of the lens, many off-axis aberrations are eliminated or minimized. The resolution and contrast are identical across every region of the specimen. The resolution and contrast are independent from the magnification; there is space-invariant imaging.
- The selection, care, cleaning, and use of a microscope objective is critical. Dirt and mechanical scratches severely degrade image formation.
- Even for imaging thin specimens, there is an increase in contrast with a confocal microscope. The confocal microscope rejects light from spots adjacent to the illuminated spot.
- The lateral and axial resolution of a confocal microscope are enhanced by a factor of 1.4 compared to the nonconfocal microscope.

Part III

Nonlinear Microscopy

Chapter 10

The Development of Nonlinear Spectroscopy and Microscopy

The introduction of the confocal microscope and its subsequent development provided an important tool for the biologist and the clinician: a means to achieve *in vivo* microscopy of cells, tissues, and organs, resulting in images with enhanced resolution, depth discrimination, and contrast. However, as researchers and clinicians pushed the limits of confocal microscopy, it became clear that new microscopes were required to image deeper into thick, highly scattering and absorbing specimens. Furthermore, as new fluorescent probes with absorption bands in the ultraviolet region were being developed for molecular biology, developmental biology, and neurobiology, there was an increasing need for confocal microscopes to operate in the UV region. In addition to the problems of obtaining microscope objectives and other optical components suitable for UV light, it was apparent that the light was toxic to living cells.

The partial solution to these increasingly important limitations came from an unexpected source: the field of nonlinear optics. Part III of the book describes and analyzes the history of nonlinear optics, the development of nonlinear microscopy, and the theory and instrumentation of multiphoton excitation microscopy.

Nonlinear optical spectroscopy preceded the development of and served as the foundation for nonlinear microscopy, which is why it is included in this textbook. Multiphoton excitation microscopy is one type of nonlinear microscopy, and so it is instructive and important to review its antecedents.*

The emphasis in this chapter is on experimental studies following the invention of the ruby laser by Theodore Maiman in 1960. While these nonlinear techniques were developed to explore the symmetry-forbidden excited states of molecules that are not realized by linear excitation, they also provided new modes of contrast in microscopic imaging. Together, these technical advances and the advent of lasers that produced femtosecond (fs) pulses made possible the invention of microscopes based on multiphoton excitation processes. Note that picosecond (ps) lasers can also be used with a multiphoton microscope, although they require much higher average power to be equally effective.

---

* Chapters 10 and 11 are condensed, edited versions of materials found in the following two sources: (1) *Selected Papers on Multiphoton Excitation Microscopy*, B. R. Masters, Ed., SPIE Press, 2003; and (2) "Antecedents of two-photon excitation laser scanning microscopy," B. R. Masters and P. T C. So, *Microscopy Research and Technique* **63**, 3–11, 2004.

Following the invention of the pulsed ruby laser in 1960, there was a series of technical advances in laser design. The advent of new types of pulsed lasers with higher peak powers and shorter pulsewidths resulted in many new and important developments in nonlinear spectroscopy and microscopy.

Again and again we observe the problem of technology transfer between different fields of study. Many technical advances were made in the fields of molecular spectroscopy and the microscopy of crystals and semiconductor devices. Cell biologists interested in the microscopic imaging of cells and tissues neither read the articles in the appropriate physics and engineering journals nor attended meetings at which this work was disseminated.

Instead, they were modifying standard commercial scanning laser confocal microscopes in order to provide laser excitation to chromophores with absorption bands in the UV region. Confocal microscopes with ultraviolet excitation resulted in new technical problems: researchers observed the rapid photobleaching of the chromophores, the poor penetration of the light into thick tissue, and, more important, the deleterious effects of the light on living cells and tissues. None of these observations was new—the history of wide-field fluorescence microscopy using ultraviolet light is replete with descriptions of these problems.

One wonders how many cell biologists asked whether there was a nonlinear spectroscopic technique that could be the basis of a new type of nonlinear microscopy. In 1929, a doctoral student in Göttingen, Germany, derived the theoretical foundation of two-photon spectroscopy. Following the invention of the laser, between 1960 and 1990, nonlinear spectroscopy rapidly expanded as a field of research. At Oxford University, the group of Sheppard and his coworkers constructed scanning laser microscopes that were based on nonlinear spectroscopy. They suggested—but did not construct—a nonlinear scanning microscope based on two-photon excitation. Was anyone in the biological community paying attention? Apparently someone was, because in 1990, at Cornell University, the group of Denk, Strickler, and Webb invented the two-photon laser-scanning fluorescence microscope. And that seminal invention forever changed the way cells and tissues are imaged!

10.1 Nonlinear Optical Processes in Spectroscopy and Microscopy

What do we mean by nonlinear optical processes? Light can interact with matter in a linear manner; hence, the effect is proportional to the intensity; e.g., single-photon absorption and fluorescence. When the light is extremely bright, the interaction can become nonlinear; that is, the optical effect is proportional to the square or the cube of the light intensity. Thus, the very strong electric fields associated with the intense light can nonlinearly alter the optical properties of the matter during the interaction.

Following the definition of Boyd in his book *Nonlinear Optics* (Boyd, 2003), **nonlinear optical phenomena** are experimentally defined when the interaction between light and materials depends in a nonlinear manner on the strength of the optical field. In linear optics, the induced polarization of matter is a linear function of the electric field strength. In second-order processes, the induced polarization is a function of the second-power of the field strength; in a third-order process, the induced polarization is a function of the third-power of the field strength. For a deeper insight into nonlinear optics, the reader is directed to the second edition of Boyd's textbook.

In general, a bright light source is necessary to detect nonlinear optical effects. When the medium has a quadratic relationship between the polarization density and the electric field, there are second-order nonlinear optical phenomena; i.e., second-harmonic generation (SHG) with the frequency doubling of monochromatic light, and frequency conversion with the sum or difference of two monochromatic waves giving the frequency of a third wave. Similarly, a third-order relationship between the polarization density of the medium and the electric field gives rise to third-harmonic generation, self-focusing, optical amplification, and optical phase conjugation.

The group of Franken in 1961 made the first observation of SHG in a quartz crystal irradiated with a ruby laser. The authors stated that the pulsed ruby laser that outputs monochromatic light (6943 A) can be focused and exhibit electric fields of 10^5 V/cm. When this light is focused onto a quartz crystal, the second-harmonic signal at 3472 A is produced. Bloembergen, in his paper "Nonlinear optics: past, present, and future," (Bloembergen, 2000) related an interesting historical aspect of scientific publication: Franken et al. passed a ruby laser pulse through a quartz crystal. They used a monochromator and a spectrographic plate to detect UV light from the experiment. However, since the spot of the second harmonic was so weak on the spectrographic plate, the editorial office of *Physical Review* removed the spots from the figure prior to publication. An erratum was never published!

Around the same time as Franken, Kaiser and Garrett made the first experimental observation of two-photon excitation fluorescence (Kaiser, Garrett, 1961). Their paper was published only a few weeks after the publication of Franken's paper, and they also used a ruby optical maser (the term "optical maser" preceded the term "laser") for their studies. They generated blue fluorescent light ~ 4250 A by illuminating crystals of CaF_2:Eu^{2+} with red light at 6943 A from a ruby laser. The authors state that the appearance of fluorescence indicated that the Eu^{2+} was in the excited state, and that the excited state was excited by a two-photon process. The authors show a log-log plot of fluorescent intensity versus incident intensity with a slope of 2, which is expected for a two-photon process.

In the following years, higher-order multiphoton processes—for example, three-photon excitation fluorescence—were observed. In 1964, Singh and Bradley reported three-photon absorption processes in naphthalene crystals with a ruby laser. The intensity of the fluorescence signal increased as the third power of the laser intensity that is expected for this nonlinear process. In 1970, Rentzepis and coworkers observed and photographed three-photon excited fluorescence in organic dye molecules.

One motivation for the intense research activity in multiphoton absorption spectroscopy is that the selection rules for electronic transitions from single-photon interactions—i.e., that two-photon absorption is an even-parity transition, and single-photon absorption is an odd-parity transition between the ground and excited electronic states—are not valid. A single-photon transition can occur between electron states with opposite parity. A two-photon electronic transition between the ground state and the excited state can couple states with the same parity. Another reason for the great interest in multiphoton processes is that two-photon absorption processes can excite higher energy states. In multiphoton spectroscopy it is possible to study Rydberg states that contain high quantum numbers. Multiphoton disassociations of molecules can explore molecular spectroscopy of high-energy states. Three-photon excitation spectroscopy is similar to single-photon excitation processes in that the electronic transitions have odd parity.

The next step in the application of nonlinear optics was the development of a nonlinear optical microscope for the examination of the microscopic structure of polycrystalline ZnSe. Hellwarth and Christensen in 1974 used a conventional harmonic microscope in which the entire specimen was wide-field illuminated with the laser beam, and the image of polycrystalline ZnSe was formed by imaging the emitted second-harmonic radiation. Their second-harmonic microscope used a repetitively Q-switched Nd:YAG laser giving 10^3 pulses per second at 1.06 μ, each of 10^{-4} J energy and 2×10^{-7} sec duration. The green second harmonic was viewed by a microscope or on a Polaroid film. Their nonlinear SHG microscope was used to observe single-crystal platelets that were not visible in an ordinary polarizing light microscope.

10.2 The Nonlinear, Scanning, Harmonic Optical Microscope is Invented at Oxford University

In 1977, the Oxford University group of Sheppard, Kompfner, Gannaway, Choudhury, and Wilson first suggested and then demonstrated how a nonlinear optical phenomenon would be incorporated into their high-resolution, three-dimensional resolved scanning laser microscope.

The principle of nonlinear scanning microscopy is explained simply in *Theory and Practice of Scanning Optical Microscopy* (Wilson and Sheppard, 1984). In a conventional light microscope, the object is illuminated with full-field illumination from an extended source through a condenser lens, the illuminated patch of the specimen is imaged by the objective lens into the image plane, and then it is viewed through an eyepiece. In such a conventional microscope, the source of the contrast is the differences in the absorption coefficients and the optical thickness of the specimen.

If a high-intensity beam of light impinged on a specimen, the specimen would behave in a nonlinear manner and higher optical harmonics would be produced. This nonlinear harmonic generation would be a function of the molecular structure of the specimen.

The Oxford group suggested developing a nonlinear scanning optical microscope where the excitation light is focused to a small volume (a diffraction-limited spot) and an image is generated based on raster scanning of either the light or the specimen. Since the excitation light is focused to the diffraction limit, the electric field strength at the focal volume is significantly greater than wide-field illumination geometry of previous designs. The significantly higher field strength allows vastly more efficient generation of optical harmonics since the nth-order harmonic signal is generated in proportion to the nth power of the fundamental intensity. For example, SHG depends quadratically on the incident light intensity. Their microscope combined nonlinear optical phenomena and laser-scanning microscopy.

The Oxford research group also pointed out an important consequence of the nonlinear dependence of the emission signal with the excitation light: super-resolution. If the fundamental radiation had a Gaussian distribution, the harmonic radiation will also have a Gaussian distribution, but the radius will be only $1/\sqrt{n}$ of that of the fundamental beam. They further noted that this system will have depth discrimination because the intensity point spread function of this microscope is a quadratic function of the intensity transfer function of the objective, similar to an incoherent confocal microscope.

Wilson and Sheppard suggested that a laser microscope could be used to investigate the nonlinear optical properties of a specimen based on variations in the specimen's second-order susceptibility. In addition to harmonic generation, they realized that other nonlinear effects such as Raman scattering, sum frequency generation, or two-photon fluorescence could be used to study the energy levels and hence the chemical structure of the specimen. In 1984 they published the suggestion that two-photon fluorescence (two-photon absorption) could be implemented in a scanning microscope.

Sheppard and his coworkers made the first conference presentation of a scanning SHG microscope in 1977. They showed a schematic for a scanning harmonic microscope, and noted that the second harmonic is formed in the forward direction. The specimen was scanned relative to the focused laser beam, and the focused beam produced optical second harmonics in the specimen itself. They proposed that a second-harmonic microscope could be used to image biological structures with very high contrast. They stated the temperature rise must be small in biological samples. They calculated that 1 W of incident light may produce 10^{-10} W of second-harmonic light. To keep the specimen temperature rise low, the laser beam and not the sample must be scanned. The authors built a specimen scanning second-harmonic microscope based on a 1 W cw Nd:YAG laser; while it was unsuitable for biological specimens, it was used to image crystals. They further illustrated the optical sectioning capability of their harmonic generation microscope by imaging various planes within a thin crystal.

Gannaway and Sheppard's 1978 paper is the first journal publication on scanning SHG microscopy. It included a discussion of the advantages of a pulsed beam and heating effects. The authors stated that their ultimate aim was to have a second-harmonic microscope to examine biological specimens. They pointed out the advantage of the scanning technique in nonlinear microscopy since a much lower laser power is necessary to achieve a given power density in the specimen. They also pointed out the advantage of using pulsed lasers to enhance the conversion of the fundamental to harmonic power in a scanning optical microscope.

Later, the Oxford group proposed several microscope configurations for the enhancement of nonlinear interactions in the scanning optical microscope, such as placing the specimen in a resonant cavity and using beam pulsing (pulsed lasers) to improve the conversion efficiency from fundamental to harmonic power. They proposed various types of nonlinear interactions for use in the scanning optical microscope: the generation of sum frequencies, Raman scattering, and two-photon fluorescence.

All of these interesting developments were presented at electronics conferences and, generally speaking, this body of work was not communicated to cell biologists; thus, the long delay in the development of nonlinear microscopy for cellular imaging.

10.3 The Role of Lasers in the Development of Nonlinear Microscopy

Many experimental advances in nonlinear laser spectroscopy and later in nonlinear microscopy depended on the generation of short laser pulses with very high peak power. Two important techniques that can generate laser pulses are Q-switching and mode locking. The technique of **Q-switching** results in very short laser pulses in which the peak power per pulse is many orders of magnitude higher compared to the same laser operating in a steady state. The symbol Q is a quality factor that specifies the sharpness of the frequency transmitted within the laser cavity. Q-switching involves the use of an optical device within the laser cavity that can rapidly switch from a high-loss to a low-loss laser cavity.

Several methods can induce Q-switching within the laser cavity; they rapidly change the Q of the laser cavity and result in the generation of "giant pulses" with very high peak power. The first method used with a ruby laser is a rapidly rotating mirror. Later, other active devices were developed: electro-optical shutters, Pockels cells, Kerr cells, acoustic-optical shutters. A saturable absorber placed within the laser cavity can also function as a passive Q-switching device to generate pulses. Q-switching has a limitation in that the minimum pulse durations are a few nanoseconds (10^{-9} s).

To overcome this temporal limitation, the technique of **mode-locking** was developed, which can produce pulses of duration that are a few femtoseconds (10^{-15} s). Note that in 50 fs, light travels a distance of only 15 μm. Mode-locking requires a small frequency spread of the laser light, and the bandwidth of the laser radiation is inversely related to the duration of the laser pulses. The term bandwidth refers to the spread of frequencies of the laser light; for example, the pulses from a mode-locked laser with a pulse duration of 100 fs would correspond to a bandwidth of 10^{13} Hz.

The technique of mode-locking generates laser pulses that have many modes. For example, a laser with a 1-m laser cavity would mode lock about 10^4 modes into the 100 fs pulses. The basis of the technique is the in-phase combination of several longitudinal modes of the laser, where each mode is at a slightly different frequency, which results in a pulsed laser output.

One passive mode-locking technique used to produce ultrashort laser pulses is called **colliding-pulse mode-locking**. It was used to form the first experimental femtosecond source for multiphoton excitation microscopy. A ring laser for colliding-pulse mode locking has two counterpropagating pulses that interact in a small spot of a saturable absorber. The formation of ultrashort pulses involves bleaching and interference within this small spot. The net result is the formation of femtosecond laser pulses. Most modern femtosecond lasers use a Kerr lens mode-locking system. A cw laser, such as an argon laser, pumps a crystal such as Ti:Al_2O_3; the Ti:sapphire laser outputs femtosecond pulses of very high peak power.

In 1961, Q-switching was used to obtain short laser pulses with high peak intensities. Laser pulses of picosecond duration were obtained by passive mode-locking with a saturable dye cell. In fact, picosecond laser pulses have been used to induce two-photon absorption in organic molecules.

In the 1970s it was possible to obtain femtosecond laser pulses based on a combination of saturable gain in a dye laser medium and a saturable dye absorber in a ring laser cavity mode together with compensation of the dispersion in group velocity. In the 1990s the development of the Ti:sapphire femtosecond laser by Spense and coworkers resulted in a laser source that was optimal for two-photon excitation microscopy. The Ti:sapphire lasers are self-mode-locked and use the Kerr lens effect to generate mode-locked pulses with output pulsewidths in the femtosecond range. This is because of the electric field generated by a strongly focused Gaussian laser beam, which causes an inhomogeneous change in the refractive index of the Ti:sapphire crystal. This change creates a weak lens in the crystal that results in a higher gain for mode-locked laser pulses than for cw pulses.

Titanium-sapphire lasers offer a broad tunability that spans the range of 700–1100 nm. The lasers can be pumped with either an argon-ion laser or a solid state semiconductor laser. The pulses are generated by the self-mode-locking Kerr effect in the lasing rod. Titanium-sapphire laser systems are most commonly used for multiphoton excitation microscopy, providing high average power (1–2 W), high repetition rate (80–100 MHz), and short pulse width (80–150 fs).

A recent development has been the use of all-solid state pump lasers, accomplished by using an array of semiconductor lasers exciting Nd:YVO_4 crystals. These lasers have the following characteristics: wavelength 532 nm, average power 10 W, single longitudinal mode with a line width < 5 MHz and a coherence length of meters, and a power stability of 1%; the output is TEM_{00} and the polarization is horizontal. This new type of laser eliminates the need for the less efficient gas lasers to pump the Ti:sapphire laser.

The next chapter introduces the work of Maria Göppert-Mayer, who in 1931 developed the theory of two-photon electronic transitions in both absorption and emission processes, which was put into practice several decades later. In 1990, Denk, Strickler, and Webb invented the multiphoton excitation microscope, and rapidly convinced the biological community of the numerous advantages of this new nonlinear microscope as compared with confocal microscopy.

10.4 Summary

- The invention of the ruby laser in 1960 and the subsequent invention of new types of pulsed lasers with higher peak powers and shorter pulsewidths resulted in major advances in nonlinear spectroscopy and microscopy.
- Many technical advances were made in and applied to nonlinear microscopy of crystals and semiconductor devices. Cell biologists neither read the articles in the appropriate physics and engineering journals nor attended meetings at which this work was disseminated.
- In 1977, the Oxford research group of Sheppard, Kompfner, Gannaway, Choudhury, and Wilson first suggested and then demonstrated how a nonlinear optical phenomenon could be incorporated into their high-resolution, three-dimensional resolved scanning laser microscope.
- Wilson and Sheppard further suggested that a laser microscope could be used to investigate the nonlinear optical properties of a specimen based on variations in the specimen's second-order susceptibility. In addition to harmonic generation, they further realized that other nonlinear effects such as Raman scattering, sum frequency generation, or two-photon fluorescence could be used to study the energy levels and hence the chemical structure of the specimen.
- The Oxford research group also pointed out an important consequence of the nonlinear dependence of the emission signal with the excitation light: super-resolution. They predicted improved transverse and axial resolution in a nonlinear microscope.
- The Oxford research group also suggested that the conversion efficiency of the fundamental to harmonic power would be increased by pulsing the laser.
- Q-switching (nanosecond pulses) and mode-locking (femtosecond pulses) are two techniques to produce short laser pulses with very high peak power.
- In the 1990s, the development of the Ti:sapphire femtosecond laser resulted in a laser source that was optimal for two-photon excitation microscopy.

Chapter 11

Multiphoton Excitation Microscopy

The history of science, in particular the field of physics, contains examples of a theoretical development preceding its experimental verification. This situation repeatedly occurred in nonlinear spectroscopy, as the experimental measurement of electronic transitions that occurs in nonlinear processes required high-intensity sources of radiation.

In this chapter I describe the long developmental path from the 1929 publication of Maria Göppert-Mayer on the theory of two-photon absorption and emission in atoms to the 1990 publication of Denk, Strickler, and Webb that demonstrated two-photon microscopy. The bridge between the work of Göppert-Mayer and the experimental realization in 1990 was the work and publications of the group at Oxford working on nonlinear scanning optical microscopy.

11.1 Göppert-Mayer's Theory of Two-Photon Absorption

The theoretical basis for two-photon quantum transitions (absorption and emission) in atoms was the subject of a doctoral thesis published in 1931 by Maria Göppert-Mayer (see Fig. 11.1). Two years earlier she published a preliminary paper on her theory (Göppert-Mayer, 1929), in which she formulated energy-state diagrams for both two-photon emission and two-photon absorption processes. She indicated the presence of virtual states, and she concluded that the probability for the two-photon absorption process is proportional to the square of the light intensity.

In her 1931 dissertation, Göppert-Mayer followed the technique of Dirac for the use of perturbation theory to solve the quantum-mechanical equations for the processes of absorption, emission, and dispersion of light in single photon–atom interactions. The transition probability of a two-photon electronic process was derived by using second-order, time-dependent perturbation theory. Her derivation clearly states that the probability of a two-photon absorption process is quadratically related to the excitation light intensity. For readers who cannot read the 1931 dissertation in German, I have made a translation into English. This translation will appear as a chapter in the new book to be published by Oxford University Press: *Handbook of Biological Nonlinear Microscopy* (Masters, So, 2006).

An important aspect of Göppert-Mayer's work is that the process of two-photon absorption involves the interaction of two photons and an atom. This interaction must occur within the lifetime of an intermediate **virtual state**, which can be described as a superposition of states and not an eigenstate of the atom. Therefore,

Figure 11.1 Dr. Maria Göppert-Mayer and her daughter. (Courtesy AIP Emilio Segrè Visual Archives.)

the probability of the two-photon transition has contributions from all intermediate states. The first photon induces the transition from the ground state to the virtual state, and the second photon induces the transition from the virtual state to the excited state. Both photons interact to induce the transition from the ground state to the excited state (see Fig. 11.2). Since the probability of the two-photon absorption processes is very low, it is necessary to use a high-intensity light in order to achieve a measurable effect; i.e., 10^{20} to 10^{30} photons per cm^2 s.

In respect for the work of Göppert-Mayer, the units of a two-photon absorption cross section are measured in GM (Göppert-Mayer) units. One GM unit is equal to 10^{-50} cm^4 s/photon. Note that a two-photon absorption cross section is not an area—the two-photon cross section does not have squared units of length as do most cross sections.

11.2 The Denk, Strickler, and Webb 1990 *Science* Publication and 1991 Patent

This section illustrates the advantage of reading the patent literature. In many cases the intellectual motivation for the invention as well as many important details of the invention are only revealed in the original patents. Perusal of both the 1990 *Science*

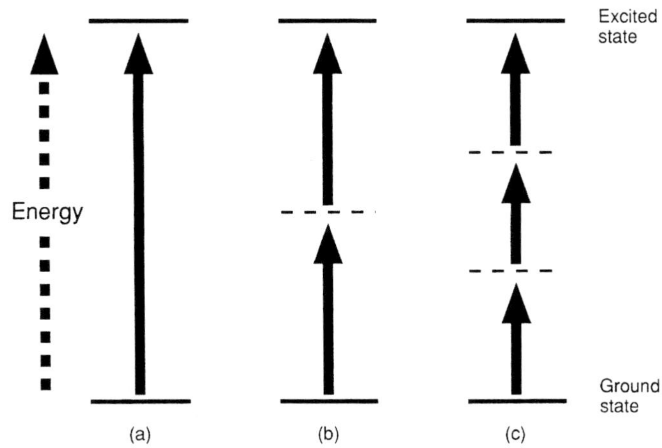

Figure 11.2 Schematic diagram showing the absorption processes for a two-level molecule with (a) one-photon absorption, (b) two-photon absorption, and (c) three-photon absorption. The dashed lines represent virtual states.

paper and the 1991 patent serves to confirm the previous statement. Both the patent and the *Science* paper are reproduced in full (Masters, 1996). While some of the concepts and principles have been previously discussed, it is of value to review the details of the invention as stated in the words of the inventors.

We now explore in more detail the invention of the multiphoton excitation microscope. As we have reviewed, it was Sheppard and Kompfner who in 1978 proposed several modes of nonlinear optical microscopy. They suggested both SHG microscopy and two-photon fluorescence microscopy. Even so, they did not construct a two-photon excitation microscope.

It was the seminal work of Denk, Strickler, and Webb, published in *Science* in 1990, that launched a new revolution in nonlinear optical microscopy and its biological applications. On July 23, 1991, they received a U.S. Patent on "Two-photon laser microscopy." The *Science* paper contains figures and text that is elaborated in their patent. The following discussion follows from their patent.

By integrating a laser-scanning microscope (scanning mirrors, PMT detection system) and a mode-locked laser that generates pulses of near-infrared light, they succeeded in demonstrating a new type of fluorescent microscope based on two-photon excitation of molecules. The pulses of red or near-infrared light (700 nm) were less than 100 fs in duration, and the laser repetition rate was about 100 MHz. The patent states that "focused subpicosecond pulses of laser light" are used. These pulses have sufficiently high peak power to achieve two-photon excitation at reasonable rates at an average power less than 25 mW, which is tolerable to biological samples. As we have previously learned, two-photon excitation microscopy can also be implemented with a picosecond laser, which was, however, not covered in their patent! As early as 1972, picosecond lasers were used for two-photon absorption studies and measurement of absorption cross sections by Bradley.

The high-intensity short pulses of near-infrared light cause significant multiphoton excitations; furthermore, the relative transparency of cells and tissues for infrared radiation and the lower average power minimizes photo-damage.

The benefits of two-photon excitation microscopy include improved background discrimination, reduced photobleaching of the fluorophores, and minimal photodamage to living cell specimens. The inventors proposed the application of two-photon excitation microscopy for optical sectioning three-dimensional microscopy and for uncaging of molecules inside cells and tissues.

The patent can be summarized by the following sentence from the Abstract, "A laser-scanning microscope produces molecular excitation in a target material by [the] simultaneous adsorption of two photons to thereby provide intrinsic three-dimensional resolution." The patent also states, "the focused pulses also provide three-dimensional spatially resolved photochemistry that is particularly useful in photolytic release of caged effector molecules."

The patent gives the light source as: "strongly focused subpicosecond pulses of laser light." The strong focusing occurs only in the focal region of the microscope objective and is similar to the origin of the optical sectioning in the second-harmonic microscope described by Wilson and Sheppard in 1979. The laser-scanning microscope described in the patent is similar to instruments described by others in prior art (the patent cited 14 previous laser-scanning instruments).

In the section of the patent labeled "Background of the invention," the authors review the various types of confocal microscopes, their light sources, and scanning mechanisms. The authors clearly state the limitations of confocal microscopy as applied to fluorescent molecules that are excited in the ultraviolet: (1) the lack of suitable microscope objectives for the ultraviolet that are chromatically corrected and transparent for both the absorption and emission wavelengths, (2) photodamage to living cells by the ultraviolet light, and (3) the problem of photobleaching of the fluorophores.

In the section labeled "Summary of the invention," the authors propose that their invention overcomes these difficulties. The authors state that the two-photon excitation is made possible by (a) a very high, local, instantaneous intensity provided by the tight focusing of the laser-scanning microscope in which the Gaussian laser beam is focused to a diffraction-limited waist of less than 1 μm, and (b) the temporal compression of the pulsed laser. This process yields improved background discrimination, reduced photobleaching, and minimizes the photodamage to living specimens.

The physics of the process is clearly described by the authors in the following sentence from their patent: "only in the region of the focal point on the object plane at the waist formed by the converging and diverging cones is the intensity sufficiently high to produce two-photon absorption in the specimen fluorophore, and this intensity dependence enables long wavelength light to provide the effect of short wavelength excitation only in the small local volume of the specimen surrounding the focal point."

The patent further provides a formula for the number of photons absorbed per molecule of fluorophore per pulse as a function of pulse duration, repetition rate, average power of the incident laser, the NA of the focusing lens and the photon absorption cross section. In a two-photon excitation process, the number of photons absorbed per molecule of fluorophore per pulse scales with the average incident laser power squared. This is the source of the experimental verification of two-photon excitation processes; hence, on a log-log plot of detected intensity versus laser power, the slope of the plot is 2. The authors also state that the two-photon excitation fluorescence can be increased by increasing the pulse repetition frequency until saturation of the excited state is achieved.

Another key feature of the patent is the description of a non-descanned detection of the fluorescence intensity derived from the two-photon absorption process. Since the fluorescence signal depends on the square of the excitation intensity, there is an optical sectioning effect through the specimen even in the absence of a pinhole used as a spatial filter in front of the detector. Therefore, the detector can be a large-area detector such as a PMT. This avoids many of the problems associated with conventional LSCMs. With the publication of the *Science* paper and the patent from the Webb group in 1991, the reality of two-photon excitation microscopy began.

Webb and coworkers further stated that the microscope can be operated in sum or difference frequency mode. It is not necessary that the two photons that are absorbed in a two-photon excitation process be of the same wavelength. In the sum frequency case, an excitation transition requiring energy hc/λ_{ab} can be achieved using lasers with wavelengths, λ_a and λ_b, based on the following conservation equation:

$$\frac{1}{\lambda_{ab}} = \frac{1}{\lambda_a} + \frac{1}{\lambda_b}, \tag{11.1}$$

where h is Planck's constant, and c is the speed of light.

Both the *Science* paper and the patent had a great impact on scientists worldwide. As soon as a commercial version was available scientists scrambled to apply this new form of microscopy to live cell microscopy.

11.3 Comparison of Multiphoton Excitation Microscopy and Confocal Microscopy

In a confocal microscope, the confocal aperture or pinhole location is critical. Also, the position of the detector aperture must be precisely located within a plane that is conjugate with the image plane and therefore the focal plane, so that the images of the source and detector apertures are cofocused.

However, in a multiphoton excitation microscope the position of the detector is not critical. The fluorescence need not be descanned prior to detection and a detector pinhole is not necessary. A non-descanned wide-area detector can be placed anywhere, but preferably close to the microscope objective after the dichroic beamsplitter.

The advantages of these nonlinear microscopes include improved spatial resolution without the use of pinholes or slits for spatial filtering. Since the optical sectioning capability of the two-photon excitation microscope derives from the physics of the excitation process, there is no requirement for a spatial filter in front of the detector as in confocal microscopy. Two-photon microscopy furthermore allows deeper penetration into thick, highly scattering tissues, and confines photobleaching and photodamage to the focal volume. The excitation wavelength is far removed from the fluorescence wavelength. This large separation in wavelengths makes the suppression of the laser light more efficient. Finally, a very important factor is increased cell and tissue viability with infrared illumination.

The main limitations of two-photon excitation microscopy are (1) it is only suitable for fluorescent imaging; reflected light imaging is not possible, and (2) it is not suitable for imaging highly pigmented cells and tissues that absorb near-infrared light.

Multiphoton excitation microscopy has the capacity to image deeper within highly scattering tissues such as *in vivo* human skin; its ability to image the elastin and collagen fibers within the tissue is a significant advantage.

The technique of two-photon excitation microscopy is ideal for deep-tissue imaging. This nonlinear microscopy technique has several notable characteristics: there is submicron resolution (0.3 μm lateral resolution; 0.8 μm axial resolution; a penetration depth that is tissue dependent (500 to 1000 μm); and has the capability of spectroscopic analysis of the specimen. The excitation wavelength is far removed from the fluorescence wavelength. This large difference in wavelength makes the suppression of the laser light more efficient.

The technique of single-photon confocal microscopy has the joint capability of either fluorescence confocal imaging or reflected light confocal imaging.

What have we learned about the fundamental divide between conventional (wide-field) fluorescence microscope on one hand, and confocal and multiphoton microscopes? The most important aspect is the property of optical sectioning in confocal and multiphoton microscopes; furthermore, it is the result of the PSF having a maximum at the focal plane. More specifically, the integrated signal intensity of the detection PSF is a constant as a function of depth in a conventional (wide-field) fluorescence microscope. On the contrary, and this is the most important statement about the origin of optical sectioning, for both confocal microscopes and multiphoton excitation microscopes the integrated intensities have a maximum at the focal plane! There is depth discrimination in both confocal and multiphoton excitation microscopes. As we shall see in the next paragraph, the widths at half height of the peaks are different.

Theoretical modeling of the point spread function for an ideal confocal microscope versus a multiphoton microscope using the same fluorescent molecule indicates a broadening in x, y, and z for the multiphoton microscope. That is because of the doubled wavelength for the multiphoton excitation microscope. This small difference in resolution can be eliminated by the use of a confocal aperture in front of the detector of the multiphoton excitation microscope. There is also a loss of signal with confocal detection. In the case of a real confocal microscope, the resolution is degraded by chromatic aberration, the use of a finite confocal aperture for efficient detection, and possibly imperfect alignment. Therefore, in practice, for imaging the same fluorescent molecule there may be no significant difference in resolution between a confocal microscope and a multiphoton excitation microscope.

When we compare the optical sectioning capabilities of both confocal and multiphoton microscopes we notice a difference; the optical sectioning capability of the multiphoton microscope is less than that for a confocal microscope. The explanation is as follows. The wavelength of the illumination in a multiphoton microscope is longer (usually by a factor of 2) than that used with a confocal microscope. While the integrated PSF for both the confocal microscope and the multiphoton microscope have peaks that occur at the focal plane, because of the longer wavelength of the illumination in the multiphoton microscope, its PSF peak is wider than the corresponding PSF peak for the confocal microscope.

For both types of microscopes, the signal rapidly falls to zero outside the focal plane. Thus, with defocus the signal disappears. Remember that for a conventional (wide-field) fluorescence microscope the signal is constant with defocus. This property of depth discrimination is key to three-dimensional microscopy via optical sectioning of a thick specimen.

Confocal microscopes use spatial filtering to obtain optical sectioning (depth discrimination). A small pinhole will achieve good depth discrimination and background rejection.

Multiphoton excitation microscopes use the physics of the excitation process, in which the excitation is proportional to the square or cube of the intensity and only within the diffraction-limited focal volume of the microscope objective is there sufficient intensity for the process to occur, to obtain optical sectioning. Multiphoton excitation microscopes have the advantage of using non-descanned detection, therefore a two-dimensional solid state detector with a very high quantum efficiency (e.g., 90%) can be used to maximize the signal.

An important advantage of multiphoton excitation microscopy is that photobleaching is restricted to the focal volume (see Fig. 11.3). In a confocal mi-

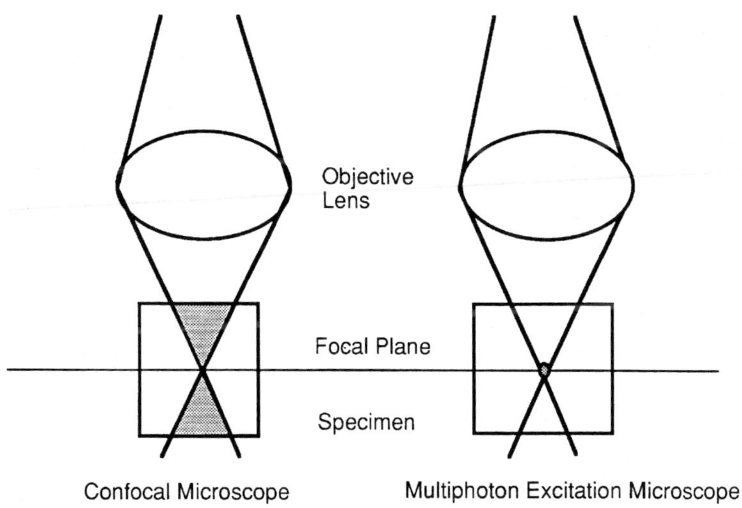

Figure 11.3 Schematic diagram comparing the photobleaching zones for a one-photon confocal microscope and a multiphoton microscope.

croscope, photobleaching occurs in all regions of the double, inverted cone of the excitation light from the microscope objective.

In the next section I describe and compare the theory and the components of multiphoton excitation microscopy. The emphasis is placed on those characteristics that optimize their use with multiphoton excitation microscopes.

11.4 Summary

- The theoretical basis for two-photon quantum transition (emission and absorption) in atoms was the subject of a doctoral thesis published in 1931 by Maria Göppert-Mayer.
- Göppert-Mayer showed that the probability for the two-photon absorption process is proportional to the square of the light intensity.
- The seminal work of Denk, Strickler, and Webb launched a new revolution in multiphoton excitation microscopy and its application to biology.
- An advantage of multiphoton excitation microscopy is improved spatial resolution without the use of pinholes or slits for spatial filtering as in confocal microscopy. The optical sectioning (depth discrimination) capability of two-photon excitation microscopes derives from the physics of the excitation process. Only in the focal volume is the intensity sufficient for observable two-photon absorption to occur.
- The use of infrared or near-infrared illumination in multiphoton excitation microscopy results in other important advantages, including greater sample penetration and increased cell and tissue viability. These factors are extremely important for live cell and tissue imaging over extended periods of time, and for applications of *in vivo* microscopy to optical biopsy.
- The main limitations of two-photon excitation microscopy are (1) it is only suitable for fluorescent imaging and not reflected light imaging; and (2) it is not suitable for imaging highly pigmented cells and tissues that absorb near-infrared light with resulting photodamage.
- The technique of two-photon excitation microscopy is ideal for deep-tissue imaging. This technique has several notable characteristics: there is submicron resolution (0.3 μm lateral resolution; 0.8 μm axial resolution); penetration depth is tissue dependent (500 to 1000 μm); and it has the capability of spectroscopically analyzing the specimen.
- In multiphoton excitation microscopy, photobleaching is restricted to the region of the focal volume. In one-photon confocal microscopy, photobleaching occurs in the volume of the double, inverted cone of the excitation light.

Chapter 12

Theory and Instrumentation of Multiphoton Excitation Microscopy

12.1 Theory

Two-photon excitation microscopy is typically associated with the following characteristics: (1) the wavelength for two-photon absorption is typically (though not always) twice the wavelength for single-photon excitation; (2) the penetration depth of the excitation light is considerably greater since the longer-wavelength light shows less scatter and therefore can penetrate deeper into highly scattering specimens; (3) the longer-wavelength near-infrared light is less damaging to live cells and tissues; however, specimens with high absorption coefficients in the infrared can still show thermal damage; (4) the axial optical sectioning capability is intrinsic to the physics of the two-photon absorption process; hence, confocal pinholes are not required; (5) descanning of the emission is not required; and (6) a high SNR can be achieved since the excitation and emission wavelengths are widely separated.

As shown in the foregoing chapter, the two-photon process occurs through a virtual intermediate state. What is not mentioned is the process in which the first photon excites the molecule into a real, intermediate state, and the second photon excites the molecule from the intermediate state to the final excited state. This process is resonant two-photon excitation.

The physics of the two-photon excitation process leads to some extremely useful consequences. The probability of the electronic transition depends on the square of the instantaneous light intensity; this quadratic dependence follows from the requirement that the fluorophore must simultaneously (within the lifetime of the virtual state) absorb two photons per excitation process.

The laser light in a two-photon excitation microscope is focused by the microscope objective to a focal volume. Only in this volume is there sufficient intensity to generate appreciable excitation. The low photon flux outside the volume results in a negligible amount of fluorescence signal. In summary, the optical sectioning (depth discrimination) capability of a two-photon excitation microscope originates from the nonlinear quadratic dependence of the excitation process and the strong focusing capability of the microscope objective.

Most biological specimens are relatively transparent to near-infrared light. The focusing of the microscope objective results in two-photon excitation of ultraviolet-absorbing fluorochromes in a small focal volume. It is possible to move the focused volume through the thickness of the sample and thus achieve optical sectioning in three dimensions.

Rayleigh scattering of light by small particles within the specimen is proportional to the inverse fourth power of the wavelength of the scattered light. With multiphoton excitation microscopy the near-infrared light is scattered much less than the visible light that is used in confocal microscopy. That is a partial explanation for the increased penetration into the specimen with multiphoton excitation microscopy as compared with confocal microscopy.

It is important to stress that the optical sectioning in a two-photon excitation microscope occurs during the excitation process. The emitted fluorescence can then be detected, without the requirement of descanning, by placing an external photon detection device as close as possible to the sample. There is no valid reason to descan the fluorescence, since this results in the loss of signal because of the mirrors and other optical components associated with the descanning system. It is strongly recommended that an external photon detector that has high quantum efficiency in the range of the fluorescence be situated near the sample, with a minimum number of optical components in the light path.

A key element of multiphoton excitation microscopy is localized excitation of the specimen. For the linear case of single-photon excitation as used in confocal fluorescence microscopy, as the distance from the focus (z) increases, the fluorescence decreases as z^2. On the other hand, for the two-photon excitation process, the fluorescence falls off as z^4. Therefore, almost all of the fluorescence comes from the focal volume.

Multiphoton excitation microscopy is an extension of two-photon excitation microscopy. In two-photon excitation microscopy the excitation depends on the intensity squared. In **three-photon excitation microscopy** the excitation depends on the intensity cubed. For a three-photon excitation process, three photons would interact with the fluorescent molecule within the lifetime of the virtual state to induce an electronic transition from the ground state to the excited state. The fluorescence is the result of the electronic transition from the excited state to the ground state of the molecule. An example of three-photon excitation microscopy is the three-photon excitation of diphenylhexatriene (DPH), which is a membrane probe. With excitation at 860 nm from a femtosecond laser, the observed emission spectrum of DPH is identical to that observed with a single-photon excitation at 287 nm. A plot of the log of the illumination intensity (at 860 nm) versus the log of the fluorescence intensity should have a slope of 3. Demonstration that the emission intensity depends on the cube of the laser power indicates a three-photon excitation process.

In order to experimentally demonstrate that a multiphoton excitation process is occurring, it is necessary to demonstrate the nonlinear nature of the process in the following manner. The intensity of the fluorescence is measured as a function of the intensity of the excitation light. These two measured quantities are plotted on a log-log plot, and the slopes of the linear regions of the plot are determined. A two-photon excitation process is characterized by a slope of two; a three-photon excitation process is characterized by a slope of three. This experimental verification of multiphoton excitation processes follows from the physical analysis of the processes, described below.

It is instructive to compare the expressions for the rates of one-photon and two-photon absorption processes for a single fluorophore. For a one-photon absorption process, the rate of absorption is the product of the one-photon absorption cross section and the average of the photon flux density. For a two-photon absorption process, in which two photons are simultaneously absorbed by the fluorophore, the rate of absorption is given by the product of the two-photon absorption cross section and the average squared photon flux density.

Following the extensive work of Chris Xu, it is instructive to compare estimates of multiphoton cross sections with one-photon absorption cross sections. For a virtual state with a lifetime of 10^{-15} s and simultaneous multiphoton excitation, the following results are obtained: one-photon absorption cross section of a molecule is 10^{-16} to 10^{-17} cm^2, a two-photon excitation is approximately 10^{-49} cm^2 (photons/cm^2 s)$^{-1}$, and for a three-photon excitation, the cross section is approximately 10^{-83} cm^2 (photons/cm^2 s)$^{-2}$.

In practical terms this means that the extremely small cross section for a two-photon excitation as compared to a one-photon excitation results in an extremely low probability of two-photon excitation with focused light from conventional sources. With highly focused pulsed lasers, the two-photon excitation process results in appreciable excitation.

The rate of two-photon excitation can be described analytically as shown in Eq. (12.1). This rate is expressed as the number of photons absorbed per fluorophore per pulse (n_a), and is a function of the pulse duration (τ_p), the repulse repetition rate (f_p), the photon absorption cross section (δ), and the NA of the microscope objective that focuses the light. The derivation of this equation assumes negligible saturation of the fluorophore and that the paraxial approximation is valid.

$$n_a \approx \frac{p_0^2 \delta}{\tau_p f_p^2} \left[\frac{\pi (\mathrm{NA})^2}{hc\lambda} \right]^2, \tag{12.1}$$

where p_0 is the average incident power, h is Planck's constant, c is the speed of light, and λ is the wavelength. The expression shows that the probability of two-photon absorption increases when the pulse duration is decreased, and when the laser repetition rate is increased.

12.2 Instrumentation

The cautions cited with respect to the use of confocal microscopes are also important for multiphoton excitation microscopes. The microscope and the lasers should be mounted on a vibration isolation table, and the effects of dust and dirt as well as the deleterious effects of mechanical scratches on the objective should be avoided.

There is also a danger from a focused laser beam striking an optical component, e.g., the objective, lenses, or dichroic mirrors. The extremely high power delivered to a small spot on the optical element will cause catastrophic damage.

12.2.1 Laser sources

In contrast to ordinary light—for example from an arc lamp—laser light is extremely intense, directional, monochromatic, and coherent. As the laser light emerges from a laser it diverges slightly; even so, a lens can focus the laser light to a diffraction-limited point of light of very high intensity. Both the monochromaticity of the laser light and its coherence properties (temporal and spatial coherence) are a consequence of the properties of the resonant cavity that is a physical part of the laser.

Femtosecond-pulse lasers are the most common light sources for multiphoton excitation microscopy. The femtosecond pulses have extremely high peak or instantaneous power, in the range of kilowatts to megawatts; however, the average power is in the range of milliwatts to watts. Therefore, the specimen is exposed to milliwatts of average power, yet the extremely intense peak power results in efficient multiphoton excitation of the fluorophores.

Two-photon excitation efficiency is a function of several characteristics of the pulsed laser: the average power of the laser (W), the repetition rate of the laser, and the laser pulse width. The two-photon fluorescence intensity is related to these parameters as

$$I \approx \frac{(p_0)^2}{\left(f_p \tau_p\right)}. \tag{12.2}$$

Note that the measured laser pulse width at the output of the laser can be significantly broadened by the microscope objective and other optical elements. The reason for the pulse broadening is due to dispersion, which is discussed in Sec. 12.2.3.

One technique to increase the intensity of the two-photon excitation fluorescence is to decrease the laser pulse repetition rate and simultaneously maintain the average laser power. An optical parametric amplifier can be used to achieve lower pulse repetition rates while maintaining the average power.

It is possible to induce two-photon excitation with a cw laser. Even so, as compared with a femtosecond pulsed laser, the two-photon absorption rate with a cw laser is very inefficient. In order to have the same two-photon absorption rates with both lasers, the cw laser would produce about 200 times more average power!

It is important to match the experimental requirements with other considerations such as cost, tunability, and ease of use. The laser should have the following general characteristics:

- Pulse duration: While picosecond lasers can be used for multiphoton excitation microscopy, in order to obtain optimal performance of the excitation process, a pulse duration of 100 fs is desirable.
- Pulse repetition rate: The suggested pulse repetition rate is in the range of 100–800 MHz.
- Peak power per pulse: In order to achieve good SNR at rapid scan rates, the peak power should be approximately 10 kW.

- Laser tunability: In order to cover the absorption bands of many chromophores, the laser tunability should span about 700 nm to 1100 nm.
- The laser output should be a Gaussian beam with a power stability of less than 1%.
- The laser should be easy to operate, easy to maintain, and of moderate cost ($50,000).

Laser product development is an active area, and the latest information from commercial sources can be found at the websites listed in the appendix. The argon-ion pumped Ti:sapphire (titanium-doped sapphire) laser, while expensive and difficult to operate and maintain, yields the highest peak power—it can be extended to include wavelengths from 350 to 500 nm by SHG—and greatest range of tunability—typically 700 to 1000 nm—an output of 1.5 W pulsed, a pulse repetition rate of 76 MHz, a pulsewidth of 150 fs, a noise level of 0.1%, a power stability of 3%, a beam diameter of 0.8 mm, a TEM_{00} spatial mode, and horizontal polarization.

An important advance in femtosecond laser light sources for multiphoton excitation microscopy is the development of femotosecond pulsed lasers that are pumped directly by laser diodes. This advance removes the requirement for large, expensive argon-ion lasers that require water cooling, expensive maintenance, and frequent adjustments. Diode-pumped Cr:LiSAF (chromium-doped lithium strontium aluminum fluoride) lasers are commercially available, although they have less power and more limited tunability than the argon-ion-pumped Ti:sapphire lasers. Typical laser specifications for these diode-pumped Ti:sapphire lasers are output power > 1W, tuning range 720 to 930 nm, pulsewidth < 140 fs, noise < 0.15%, beam diameter 1.2 mm, pulse repetition rate 90 MHz, and a horizontal polarization. These lasers usually include a spectrometer that shows both the wavelength and bandwidth.

Many important biological fluorophores have two-photon absorption bands in the wavelength range from 500-700 nm, including Indo-1, NADH, DAPI, and Dansyl. Tryptophan, dopamine, and serotonin can be excited with three-photon excitation in this shorter wavelength range. One laser source is the Cr:forsterite laser, with an output in the range of 1150 to 1360 nm. With SHG, this laser would yield pulses ranging between 575 nm and 680 nm and be a very useful laser source for two-photon excitation microscopy.

12.2.2 Laser beam diagnostic instrumentation

It is critical to set the optical parameters of the laser, including wavelength, average power, peak power, pulsewidth, and the pulse repetition rate, and to monitor their values. As previously discussed, the rate of multiphoton excitation is dependent on these parameters, and for optimal signal strength it is critical to select these parameters correctly, based on knowledge of the multiphoton absorption spectra and other properties of the fluorescent probe molecule, as well as sensitivity to photo-damage from the laser light.

The wavelength can be measured with a calibrated grating spectrometer. The laser pulsewidth can be measured with an autocorrelator. An autocorrelator does not directly yield the laser pulsewidth. It is first necessary to assume a pulse shape (either a Gaussian or Lorentzian pulse shape). The output signal from the autocorrelator is then deconvoluted assuming a given pulse shape, and the pulsewidth is obtained.

The average power of the laser beam is directly measured with a power or energy meter. An analog output meter is suggested for fine-tuning the power and checking the laser alignment.

The peak power of the laser is a calculated parameter. First the pulsewidth, the average power, and the pulse repetition rate of the laser are determined. Then, the peak power is calculated by dividing the average power by the product of the pulse repetition rate and the pulsewidth.

Below are representative pulsed laser parameters based on the following assumptions and values: 10 mW average laser power, 100 MHz pulse repetition rate, and a laser pulsewidth of 100 fs. Energy per pulse is calculated by dividing the average power by the pulse rate, i.e., 0.1 nJ. The peak power is calculated by dividing the energy per pulse by the pulsewidth, i.e., 1000 W. The laser duty cycle is calculated from the product of the pulsewidth and the pulse rate: 0.001%. The period is defined as the time between successive pulses and is calculated as the reciprocal of the pulse rate, calculated to be 10 ns.

12.2.3 Laser pulse spreading due to dispersion

Laser pulses have a pulsewidth of 10^{-13} s as they emerge from a mode-locked laser. As the short laser pulses propagate through the glass and multilayer dielectric coatings in the microscope and its objective, they are spread out in time. This effect is caused by **group velocity dispersion**. Since each individual laser pulse consists of a distribution of optical frequencies, the wave packets will propagate at different velocities as determined by their group velocities.

Why is dispersive laser pulse spreading important? From Eq. (12.1) we observe that n_a, the number of photons absorbed per fluorophore per pulse, is inversely related to τ_p, the pulse duration. Therefore, an increase in the laser pulse duration caused by group velocity dispersion results in a decrease in the number of photons absorbed per fluorophore per pulse. The net effect is a decrease in the fluorescence due to multiphoton excitation.

Certain pulse compression techniques, also called "prechirping," can be used to compensate for group velocity dispersion. It is possible to compensate for the spreading of the femtosecond laser pulses by dispersive optical elements. This effect results from the positive dispersion of most optical elements in the beam path of the microscope. The trick is to use several prisms or gratings that cause a frequency-dependent optical path difference that results in a negative dispersion. The sequential combination of the positive dispersion caused from the optical elements in the microscope and the negative dispersion caused by the prisms compensate each other. This was first suggested by Fork et al. in 1984.

12.2.4 Microscope objectives

The selection of the appropriate microscope objectives for the solution of the imaging problem at hand is critical. Fortunately, there is a wide selection to choose from. The following optical parameters are noteworthy: magnification, NA, free working distance, and a choice of air, water-immersion or oil-immersion objectives. If the specimen must be in air, then a noncontact microscope objective is required. If the specimen can be placed in contact with water or microscope immersion oil, then the higher NA of these microscope objectives can be utilized, which translates into increased resolution and sensitivity. A long free working distance is useful when the specimen is several microns thick or if electrical or mechanical measurements are to be performed in parallel with microscopic imaging.

Modern microscope designs are based on infinity-corrected lens systems, which have a great advantage in that the focus can be changed by moving the objective, i.e., with a piezoelectric micropositioner. In addition, the placement of optical elements within the tube length has very little effect on the primary image plane.

Typically, a motor can be used to move the microscope stage for coarse adjustment. The fine adjustment of the position of the microscope objective with respect to the specimen is controlled by a piezoelectric computer-controlled micropositioning device.

12.2.5 Scanners

Multiphoton excitation microscopy differs from confocal microscopy in that the focal volume of the illumination within the specimen completely defines the position of the light that forms the image. Therefore, it is not necessary to descan the fluorescence and send it back through the same optical path as on illumination. While that process, called **descanning**, is required in confocal microscopes, it is neither necessary nor optimal in multiphoton excitation microscopy. Instead, the fluorescence is detected in a **non-descanned mode** (see Fig. 12.1).

Non-descanned detection has several advantages. The number of reflective surfaces and lenses is minimized, which decreases the loss of signal in the microscope. Since a confocal pinhole aperture is not required in multiphoton excitation microscopy, a wide-angle detector could be used. It is optimal to locate the wide-angle detector as close to the microscope objective as possible in order to maximize light collection and thus the strength of the signal. Typically, the wide-angle detector is placed next to the dichroic mirror above the microscope objective.

Although the fluorescence need not be descanned prior to detection, it is still necessary to scan the excitation laser beam. Most commercial multiphoton excitation microscopes use a point-scanning or beam-scanning system in which the laser light forms a planar image in the specimen. The scanning systems are similar to those used for confocal microscopy.

For slow beam scanning, two mirrors mounted on oscillating galvanometers can provide the raster line scanning. These two oscillating mirrors are on orthogo-

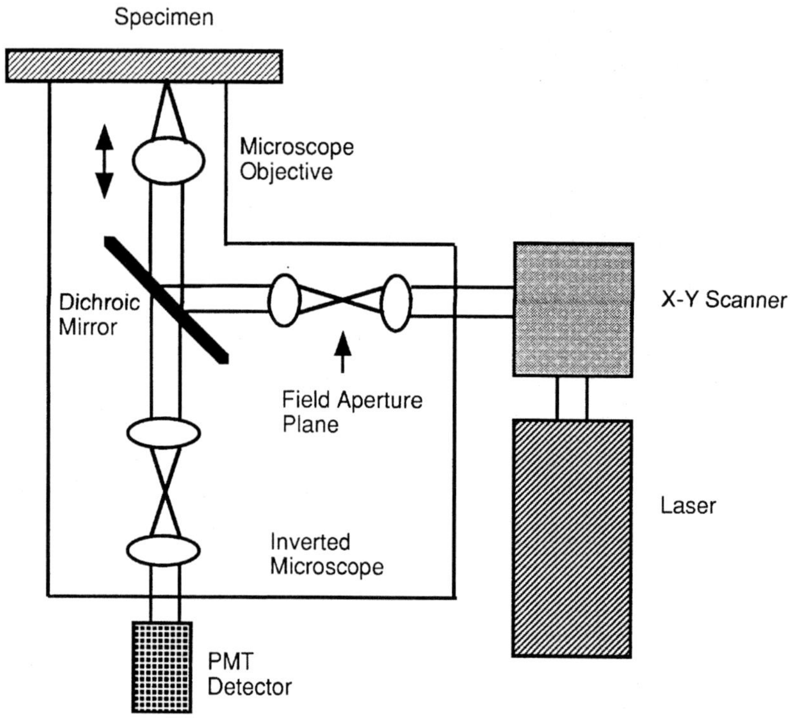

Figure 12.1 Schematic diagram of the multiphoton excitation scanning microscope. Note that there is direct detection of the fluorescence.

nal axes. When video frame rates are required with a multiphoton excitation microscope, then a polygonal mirror rotating on one axis can be combined with a single oscillating mirror mounted on an oscillating galvanometer on an orthogonal axis. This is described by Kim et al. (1999).

An alternative technique to increase the frame rate is to use a microlens array to simultaneously illuminate the back focal plane of the microscope objective with multiple beams of light. The frame rate of the multiphoton excitation microscope will increase proportionally to the number of beams in the illumination system. This is described by Bewersdorf et al. (1998). Another technique is to use a beamsplitter for multifocal multiphoton microscopy, as described by Nielsen et al. (2000).

12.2.6 Detectors

In multiphoton excitation microscopy the pinhole aperture spatial filter is not required. Nevertheless, a pinhole spatial filter can be inserted in front of the detector in order to improve the resolution. A consequence of using the spatial filter is a decrease in the signal intensity.

The choice of detectors is similar to those discussed in the section on confocal microscopy. Again, one can select a variety of photomultiplier tubes, avalanche photodiodes, and two-dimensional charge-coupled device (CCD) cameras. PMTs are most often used because they are sensitive in the blue-green region of the spectrum, are low cost, and have a high dynamic range. In addition, they can be used in the previously described photon-counting mode for detection of extremely low light intensities.

Avalanche photodiodes excel with their higher sensitivity. They have a smaller dynamic range and are more expensive. The sensitive area is smaller than for a PMT.

12.3 Summary

- In multiphoton excitation microscopy the absorption increases quadratically (two-photon absorption) or cubicly (three-photon absorption) with the excitation light intensity. The fluorescence, photobleaching, and photodamage associated with the multiphoton excitation processes are confined to the focal volume. In confocal microscopy these processes occur in the entire inverted cone of the illumination.
- The physics of the excitation process defines the optical sectioning capability.
- Detection of the fluorescence emission does not have to be descanned and the emitted light does not have to be focused. Therefore, a high-efficiency two-dimensional detector can be used. Alternatively, a PMT can be located close to the dichroic mirror.
- The longer wavelengths of laser light used in multiphoton excitation microscopy permit greater penetration depth into the specimen, and a better separation between the excitation wavelength and the emission wavelengths resulting in the possibility of detection with an increased signal to noise ratio. There is almost no background interference from Raman and Rayleigh scattering.
- In practice, the spatial resolution of the multiphoton excitation microscope is similar to that of the fluorescent confocal microscope. Multiphoton excitation microscopy has a slightly lower resolution for a given fluorescent probe compared to confocal microscopy. The use of a confocal aperture will eliminate this loss in resolution with a concomitant loss in signal.

Part IV

The Path to Imaging Live Cells, Tissues, and Organs

Chapter 13

Remaining Problems, Limitations, and Their Partial Solutions

Multiphoton excitation microscopy has several limitations. For a given fluorophore the resolution is slightly lower than with a confocal microscope. The insertion of a confocal pinhole in front of the detector can eliminate this difference, but there is a large loss of signal with the use of a pinhole.

Multiphoton excitation microscopes only work in the fluorescence mode. It is possible to collect the backscattered reflected light with a confocal pinhole in front of the detector, which provides simultaneous imaging of the specimen in both multiphoton excitation and confocal microscopy modes. The advantage of multimode imaging, with its increase of information about the object, is often very important.

The depth of imaging limitation is also very important. Depending on the nature of the specimen, nonlinear multiphoton microscopes can achieve a penetration depth that is 2 to 3 times that of a confocal microscope. The value of depth penetration is highly dependent on the concentration of absorbing and scattering molecules in the tissue under microscopic observation.

This problem has two components. The first limitation is the free working distance of the microscope objective. That is a fixed limitation of any type of microscopy and is based on the distance that the microscope objective can focus into the specimen. In an ideal, semitransparent specimen, that is the focal distance when the tip of the microscope objective just makes contact with the specimen. When a cover glass is used, this distance is reduced by the thickness of the cover glass.

The second process that limits the depth within the specimen is the amount of light scatter and absorption. Within the focal volume, multiphoton excitation processes cause the chromophore to attain higher-energy electronic states. The electronic transition from the first singlet state to the ground state is accompanied by the production of a photon with the energy of the transition. Photons that are either scattered out of the microscope objective or absorbed within the sample never are detected. If the specimen has a scattering coefficient and absorption coefficient that is constant within the depth of the specimen, then increasing depth of the specimen will scatter and absorb increasing amounts of the fluorescence; therefore, the signal will decrease.

Photobleaching of fluorescent probes is another problem. In the multiphoton excitation microscope the zone of photobleaching is constrained to the focal volume; nevertheless, the problem still exists. Perhaps the design of new fluorescent probes could mitigate this problem.

Photodamage to the specimen is a problem with both single-photon confocal microscopy and multiphoton excitation microscopy. Photo-oxidation is the mechanism of damage that occurs with two- and three-photon microscopy. The endogenous and exogenous fluorescent molecules interact with the high-intensity light and oxygen to form singlet oxygen and highly reactive free radicals, which results in cell damage and death. Two-photon as well as higher-order multiphoton processes may be involved with cellular damage and death.

A second mechanism of photodamage to the specimen may result from the extremely high peak powers of the femtosecond pulsed lasers used in multiphoton excitation microscopy. The associated high electric field strengths may cause dielectric breakdown.

A third important problem is thermal damage to the specimen. This is particularly important if multiphoton excitation microscopy is used to image skin *in vivo* or for ocular imaging. Single-photon absorption of the highly intense infrared radiation can result in thermal damage. If the specimen contains a high concentration of molecules with strong absorption bands in the infrared region, e.g., melanin in human skin, then thermal damage to the specimen can occur.

One solutions to imaging problems is to use a laser pulse picker based on an acousto-optical modulator (Masters et al., 2004). Such a device can reduce the pulse train repetition rate of the laser, though the instantaneous power of the pulses is unchanged. The average power of the light is reduced, which results in less thermal damage. The probability of the two-photon absorption process is proportional to the peak power, and that is why it should be maintained for imaging.

Much of the discussion in this book is about the development of new techniques and instruments to improve the resolution, the contrast, the background rejection, and the optical sectioning capability of the light microscope. The invention of phase contrast microscopy, differential interference contrast microscopy, confocal microscopy, and multiphoton excitation microscopy is evidence of a strong and continuing string of innovations in light microscopy.

The Abbe resolution limit for far-field light microscopy yields a limited far-field spatial resolution for the light microscope. The lateral resolution is approximately 180 nm, and the axial resolution is approximately 500 nm for visible light. These diffraction limits are valid for confocal microscopes with single-photon fluorescence imaging or with multiphoton excitation microscopes. This resolution limit has been broken, and today light microscopes can provide three-dimensional resolution in the 100-nm range. Much of this innovative technology has been developed in the laboratory of Stefan Hell in Göttingen, Germany. The appendix lists the website of his laboratory and while the details are outside the content of this book, nevertheless, it is instructive to visit that site to uncover the details of the theory and the instruments.

One technique to overcome the limited resolution that results from the finite NA of a single microscope objective is to use the coherent addition of the focused wavefronts of two opposing microscope lenses. Digital filtering is used to remove the two axial sidelobes, which improves the axial resolution of the light microscope 3 to 7 times.

Another method developed by the group of Stefan Hell is to increase the spatial resolution by decreasing the size of the diffraction-limited spot. This technique, stimulated emission depletion (STED) microscopy, enables light microscopy to exceed the diffraction limit.

Other groups have also developed a variety of optical techniques to improve the resolution of the light microscope. Ernst Stelzer and his coworkers at the EMBO laboratory in Heidelberg have developed a technique called confocal theta microscopy. Their microscope uses two microscope objectives set at different axes (about 90 deg.); one objective is used for illumination, and a second orthogonal objective is used for detection. The result is that the PSF in the x-z plane is almost isotropic. With a NA of 0.9, the full width at half height of the PSF is equal to 0.30 μm in the lateral direction and 0.28 μm in the axial direction. A more recent development from the Stelzer group is multiple imaging axis microscopy (MIAM). Four high-NA microscope objectives are used to achieve an isotropic resolution of 220 nm.

These techniques have broken the Abbe limits for the resolution of a light microscope. They represent important advances in the field of light microscopy. It is important to note that some of these new types of microscopes operate in the transmission mode.

For the biological community to accept a new microscope design, it must provide new or improved imaging characteristics. A second requirement is the degree of complexity needed to use the microscope. If we survey the development of various types of microscopes that include electron microscopes, scanning probe microscopes, and nonlinear microscopes, with the multiphoton excitation microscope as a seminal example, we see both increased performance and an improvement in the ease of operation. Hopefully, this trend will continue as new technical innovations in light microscopy are developed. The consequences will be new advances in biology and medicine and the improvement of the human condition.

Chapter 14

Speculation on Future Directions for Confocal and Multiphoton Excitation Microscopy

14.1 Correlative Microscopy

From its beginnings, the field of microscopy was plagued by artifacts. Artifacts may be thought of as observed properties in the image of microscopic objects that are not inherent to the object in its natural state. Open any histology book and there is an important section on artifacts in microscopy.

Artifacts can be placed into two classes: those associated with sample preparation which includes staining and genetically expressed fluorescent probes, and those associated with the optical system and the physics of photon detection. Examples of artifacts associated with the preparation of the sample include the following: improper sampling of a heterogeneous sample; fixation, mechanical sectioning, staining, heating and drying. Also to be included in this category are light damage to the sample, mechanical damage, thermal damage, and changes in living cells and tissues once cell death occurs.

The latter category includes optical aberrations and insufficient spatial and or temporal resolution to adequately observe structure and physiological function. For example, with inadequate transverse resolution, high spatial frequency structures will not be apparent. When monitoring physiological events, if the temporal resolution is insufficient then the true time course of these events will not be correctly measured.

Even when all due care is maintained in sample preparation and when the optical elements are selected to minimize many of the optical aberrations, the possibility of incorrect interpretation of the images still exists. That is when the power of correlative microscopy becomes apparent. As mentioned in Chapter 5, **correlative microscopy** is the use of two or more types of microscopy on the same sample. For example, a sample could be observed with both reflected light confocal microscopy and fluorescence light multiphoton excitation microscopy. Alternatively, a specimen can be observed with confocal and interference microscopy (DIC). Images obtained with optical low coherence tomography (OCT) can be compared with confocal microscopy and multiphoton excitation microscopy. When several different microscope techniques show similar structures on the same sample, then the likelihood that the images correspond to the structure of the specimen is increased.

We will see increased use of correlative microscopy and this will increase the accuracy of our observations and measurements.

14.2 Multimodal Microscopes

A multimodal microscope is one instrument that contains several imaging modes; for example, a microscope that contains both single-photon confocal microscope (reflected-light and fluorescence-light mode) capability and the multiphoton excitation microscopy mode. Another possibility is to combine interference microscopy with confocal microscopy.

Since each type of microscopy depends on different modes of contrast it would be an advantage to combine two different modes of nonlinear microscopy in order to obtain additional information about the specimen. For example, a microscope based on both SHG and two-photon excitation microscopy would be useful to investigate optical biopsy of tissues and organs. The intrinsic molecules such as NAD(P)H and flavoproteins would generate two-photon fluorescence signals with multiphoton excitation. The collagen in the tissue would contribute to the second-harmonic signal. Together, both types of nonlinear microscopy would provide new information; moreover, they combine both cellular metabolism and tissue structure in the composite images.

The advantage of multimodal microscopy is that several different modes of contrast can be used on the same specimen. This has the advantage of minimizing artifacts and increases the information content of the images. Multimodal microscopes have the advantages of correlative microscopy and in addition can be performed on the same sample.

A microscope can be constructed to have both single-photon confocal imaging and multiphoton excitation microscopy capability. In addition, it is possible to use a confocal aperture to increase the resolution of the multiphoton images. There would be a concomitant loss of signal with the aperture in front of the detector. Another version of a multiphoton microscope would pass the reflected light from the fs source through a confocal aperture in front of a detector. In that manner both the confocal reflected light image and the fluorescence image from the multiphoton excitation microscope could be simultaneously collected from the same specimen. Multimodal microscopes typically suffer from the following trade-off: in order to have several types of microscopy operating in the same instrument, it is very difficult to obtain optimal performance from each mode.

14.3 *In-Vivo* Microscopy or Live Cell and Tissue Imaging

Microscopy began with the observation of living specimens, and *in vivo* microscopy continues to show a revival. At present, we see its application to *in vivo* studies of the brain, eye, skin, developmental biology, cell biology, immunology, protein trafficking in cells, cell signaling, and optical biopsy. Further developments are resulting in new confocal endoscopes and other microscopic devices for medical imaging and diagnosis.

In the fields of ophthalmology and dermatology there are new clinical devices for diagnostics. Several types of clinical confocal microscopes exist for the clinical observation of the anterior segment of the *in vivo* human eye. Other LCSMs have been developed for the observation of the retina.

Smaller, cheaper, and more user-friendly LSCMs have been constructed for more widespread use in clinical dermatology. These LSCMs will be further enhanced with the use of eye trackers and adaptive optics; eye trackers stabilize the image of the moving eye, and adaptive optics correct for wavefront aberrations of the eye.

Another area of promising development is the incorporation of spectral imaging into *in vivo* microscopes. The ability to record the complete spectrum at each pixel dramatically increases the information that can be extracted from *in vivo* microscopy.

Two other promising developments are the use of solid state light arrays, either diode lasers or light-emitting diodes (LEDs) of various wavelengths, and the use of lifetime imaging. Both of these developments will be incorporated in the next generation of *in vivo* microscopes.

We have seen the development of several types of clinical microscopes for both ophthalmology and dermatology in the clinic. However, we have yet to see a massive technology transfer of *in vivo* microscopy to clinical diagnostic medicine. We hope this will soon change and new types of diagnostic light microscopes will emerge in the clinic.

14.4 Instrument Development

We will continue to see new laser developments in the area of femtosecond light sources for multiphoton excitation microscopy. Light sources will be developed that are cheaper and more compact, easier to maintain, and have wide ranges of tunability. We will also see the continuing development of femtosecond light sources in the range below 700 nm. Another development is the use of shaped femtosecond pulses to increase the selectivity of multiphoton excitation microscopes. Furthermore, using femtosecond pulses to pump a photonic crystal fiber will provide a visible continuum for fluorescent confocal microscopy.

Two-dimensional photon detectors are being developed with properties that impact directly on low-light-level live cell and tissue imaging. We can expect to see continuing advances in the technological development of two-dimensional CCD cameras with the following properties: high quantum efficiency (peak > 90%), extremely low noise, and a wide dynamic range that permits the detection of very bright and very weak signals in the same image.

Microscope objectives will be manufactured with longer working distances, which will permit deeper penetration into tissues and thick specimens.

A very active area of development in both confocal and multiphoton excitation microscopy is the development of new probes and contrast media. Advances in making quantum dots compatible with living cells and tissues will enhance their utility in live, long-term cell imaging, because they do not bleach as do many fluorescent probes. New techniques to introduce these biocompatible quantum dots into cells will be developed. These probes will be useful in long-term studies of cell migration, proliferation, and differentiation. In addition, the development of new fluorescent probes will continue to advance, both those that are introduced into the cells and those that are genetically expressed.

Another active area of probe development is the construction and expression of GFPs in cells, tissues, and whole organisms. These genetically expressed fluorescent proteins are having a great impact in cell biology and developmental biology. It is important to realize that these green fluorescent proteins are overexpressed and therefore the appropriate controls are required to demonstrate that these fluorescent proteins are correctly localized in the cell.

Probe and contrast media development will continue to accelerate and we will see the emergence of new types of highly specific molecular probes for *in vivo* microscopy of pathology.

14.5 Summary

- *In vivo* microscopy will continue to advance; new endoscopes will be developed.
- Instrument development will include new compact lasers and array sources of light, and spectral and lifetime imaging for *in vivo* use; many new types of microscope objectives, detectors, and types of fluorescence probes and contrast media will be developed.
- New types of diagnostic light microscopes will emerge in the clinic.
- New advances in far-field optical microscopy to image specimens up to several millimeters in thickness are being developed.

Chapter 15

Safety and Cleanliness Considerations

Two other topics that should not be overlooked are laser safety and caring for the optics in your microscope. Here is a summary of cautions to take when using lasers and light microscopes.

15.1 Laser Safety

- Read one or more source books on laser safety.
- Be aware of the different types of lasers, their damage thresholds, and the risks of ocular damage, skin damage, and fire associated with each type of laser.
- Be aware of the advantages (ocular protection) and disadvantages (no knowledge of position of the laser beam) of laser safety goggles.
- When designing a laser system on an optical table be sure that the laser beam is not at eye height.
- Be aware of the fire danger with lasers. Most materials in the clinic are flammable.

15.2 How to Clean Optics

- Optics should be cleaned at the following times: when there is a power loss; at regular maintenance intervals; when a laser cavity is opened; when a new set of mirrors is installed; when new optics are received.
- Learn how to correctly clean a microscope objective prior to and after use.
- Particles of dust can severely degrade the image quality of optical components.
- Improper cleaning of a microscope objective will cause microscopic scratches that will permanently degrade the image quality. Never use anything but lens-cleaning paper with spectroscopic-grade acetone and methanol. Note that these solvents can solubilize oil from your skin and transfer it to the surface of the optical element.
- Use a dust cover made from a conductive material to shield the microscope from dust.

Epilogue

This book is concerned with optical microscopy; the principles, instruments, and limitations of far-field microscopes. Alternative types of microscopes operate in the near-field (in which the object-to-microscope distance is less than the wavelength of light), which provide another type of optical microscopy that exceeds the classical limit of resolution (Jutamulia, 2002).

Although the emphasis of this book has been on confocal microscopes and multiphoton excitation microscopes, the reader should not fall into the trap of using those instruments that are available, familiar, or that were successful in the past. Both types of microscopes are limited by the depth of penetration and a significant difference in the lateral and axial resolution.

As research problems evolve, there is always the possibility that other types of imaging may be more appropriate and provide unexpected solutions. The great advances made in optical microscopy, both in far-field and near-field optical microscopes, have a significant impact on our visualization and understanding of the microscopic world.

Similarly, nonoptical imaging modalities such as ultrasound, computerized x-ray tomography or x-ray computed tomography (CT), and magnetic resonance imaging are in continuous development and evolution. These techniques have been adapted to a wide range of specimens, from whole body imagers to microscopes for very small specimens. The reader should be open to new and to evolving techniques. The message is simple: the instrument used in the investigation must be appropriate to the questions asked and to the specimen under observation. Resolution and contrast are partial considerations; there are also factors of safety, specimen area and thickness, and image acquisition time. Optical microscopes offer high-resolution, high-contrast images, but only for a small area.

Optical microscopy is inherently two-dimensional: the focal plane is flat. Another consideration is that the axial and lateral resolutions are very different. The depth of penetration is limited by both the microscope and the specimen. The free-working distance of the microscope objective is a limit of the instrument. The absorption and scattering coefficients of the specimen limit the light penetration into the thickness of the specimen. There are many specimens whose diameter or thickness exceeds the depth of penetration of the light microscope. Further development and progress in the fields of optical biopsy and *in vivo* microscopy for biology and medicine may reach limits based on these considerations.

Fortunately, there are new and exciting advances to solve some of these problems. Optical projection tomography has been developed to provide high-resolution three-dimensional images of both fluorescent and nonfluorescent biological specimens with a thickness up to 15 mm (Sharpe et al., 2002). Another approach to imaging large, thick, live biological specimens is selective plane illumination microscopy (Hulsken et al., 2004). This technique can generate multidimensional images of live specimens up to a few millimeters in size.

Another approach that is emerging as an important diagnostic imaging tool is optical low-coherence reflectometry. This technique was originally developed in the telecommunications industry for testing fiber optic cables and integrated optical devices. In the last decade the field has been further developed and applied to biology and diagnostic medicine (Masters, 2001).

There are still obstacles on the path to imaging live cells, tissues, and organs. The depth of penetration is not sufficient for many specimens and for "optical biopsy." The problem of phototoxicity is a major concern. Optical microscopes are inherently two-dimensional, but they are used to view a three-dimensional world.

Microscopes are tools. Humans are tool users, but they are also tool makers. In this book I have described the ingenious tools, from optical microscopes to fluorescent probes, that have been developed so we can image the invisible world of the microcosm. There is no reason to believe that this development of new tools will not continue. And with the new developments in microscopy will come increased knowledge and understanding of our world.

Appendix

Reference Materials and Resources

R. R. Alfano and B. R. Masters, Editors, *Biomedical Optical Biopsy,* Vol. 2, Classic Articles in Optics and Photonics on CD-ROM Series, Optical Society of America, Washington, D.C. (2004). This work contains introductory sections for each optical technique and reprinted papers (PDF) in the areas of linear and nonlinear optical microscopy and spectroscopy. Topics include the theory, instrumentation, and application of many optical techniques for *in vivo* microscopy.

W. Becker, *Advanced Time-Correlated Single Photon Counting Techniques*, Springer Verlag, Berlin (2005). This is an important and practical book on the topic of single-photon counting techniques. The chapter on detectors for photon counting is both clear and comprehensive and thus highly recommended.

J. Bewersdorf, R. Pick, and S. W. Hell, "Multifocal multiphoton microscopy," *Optics Letters* **23**, 655–657 (1998).

N. Bloembergen, "Nonlinear optics: past, present, and future," *IEEE J. Select. Top. Quant. Electron.* **6**, 876–880 (2000). A detailed history of nonlinear optics.

R. W. Boyd, *Nonlinear Optics*, 2nd ed., Academic Press, San Diego (2003). This is a very well written textbook that provides the theoretical foundation for modern nonlinear optics. This book provides a solid foundation to understand the fundamentals of nonlinear spectroscopy and microscopy.

S. Bradbury, et al., *RMS Dictionary of Light Microscopy*, Oxford Science Publications, Oxford University Press, Oxford, UK (1989). This dictionary defines over 1250 terms used in the field of light microscopy. The appendix gives the definitions of equivalent terms in English, French, and German.

E. M. Brumberg, "Fluorescence microscopy of biological objects using light from above," *Biophysics* **4**(4), 97–104 (1959).

A. H. Buist, M. Müller, J. Squier, and G. J. Brakenhoff, "Real-time two-photon absorption microscopy using multipoint excitation," *Journal of Microscopy* **192**(2), 217–226 (1998).

D. H. Burns, R. B. Hatangadi, and F. A. Spelman, "Scanning slit sperture confocal microscopy for three-dimensional imaging," *Scanning* **12**, 156–160 (1990).

G. Clark and F. H. Kasten, *History of Staining*, 3rd ed., Lippincott, Williams, & Wilkins, Baltimore (1983). The standard text for the history of cell and tissue staining for optical microscopy.

P. M. Conn, Editor, *Confocal Microscopy* **307,** *Methods in Enzymology*, Academic Press, San Diego, CA (1999). This is a good discussion of modern biological applications of confocal microscopy. Each application is described with full experimental details in order that the reader can use these methods. There are several chapters on *in vivo* microscopy.

T. R. Corle and G. S. Kino, *Confocal Scanning Optical Microscopy and Related Imaging Systems*, Academic Press, San Diego, CA (1996). This book is a comprehensive introduction to the field of scanning optical microscopy, including the confocal scanning optical microscope and the optical interference microscope. It contains a very clear introduction to the theory of depth and transverse resolution. This is a good source of applications in the semiconductor industry and metrology. The theory of the confocal microscope is well written.

I. J. Cox and C. J. R. Sheppard, "Digital image processing of confocal images," *Image and Vision Computing* **1**(1), 52–56 (1983).

P. Davidovits and M. D. Egger, "Scanning laser microscope for biological investigations," *Applied Optics* **10**(7), 1615–1619 (1971).

W. Denk, J. H. Strickler, and W. W. Webb, "Two-photon laser scanning fluorescence microscopy," *Science* **248**, 73–76 (1990).

A. Diaspro, Editor, *Confocal and Two-Photon Microscopy: Foundations, Applications, and Advances*, Wiley-Liss, New York (2002). This multiauthor volume is a good source for basic theory of confocal and multiphoton microscopy and many of their applications, from biology to the characterization of integrated circuits and optoelectronics.

A. Draaijer and P. M. Houpt, "A standard video-rate confocal laser-scanning reflection and fluorescence microscope," *Scanning* **10**, 139–145 (1988).

R. L. Fork, O. E. Martinez, and J. P. Gordon, "Negative dispersion using pairs of prisms," *Optics Letters* **9**(5), 150–152 (1984).

R. D. Goldman and D. L. Spector, *Live Cell Imaging, A Laboratory Manual*, Cold Spring Harbor Laboratory Press, Cold Spring Harbor, New York (2005). This is a very good resource for imaging live cells and organisms. The protocols cover mammalian cells, yeast, and tissues.

D. J. Goldstein, *Understanding the Light Microscope: A Computer-aided Introduction*, Academic Press, London (1999). This very practical book contains computer programs that allow students to simulate the effects of aperture, spherical aberration, and focus of the objective lens; the operation of bright-field and phase contrast microscopes; quantitative polarized-light microscopy; and a ray-tracing program that shows the effects of aberrations in simple and compound lenses. The book contains a good review of Abbe's elementary diffraction theory and various techniques to form contrast in optical microscopy.

D. S. Goodman, "General Principles of Geometric Optics," Chapter 1 in *Handbook of Optics*, Michael Bass, William L. Wolfe, David R. Williams, and William L. Wolfe, Editors., Optical Society of America, McGraw-Hill, 1–109 (1995). A clear introduction to geometrical optics.

M. Göppert, "Über die Wahrscheinlichkeit des Zusammenwirkens zweier Lichtquanten in einem Elementarakt," *Die Naturwissenschaften* **17**, 932 (1929).

M. Göppert-Mayer, "Über Elementarakte mit zwei Quantensprüngen," *Annalen der Physik* (Leipzig) **9**, 273–294 (1931). An English translation will appear in Masters and So, *Handbook of Biological Nonlinear Microscopy*, Oxford University Press (2006).

M. Gu, *Principles of Three-Dimensional Imaging in Confocal Microscopes*, World Scientific, Singapore (1996). This excellent book presents a clear development and analysis of the three-dimensional transfer functions for various confocal microscopes. It describes single-photon confocal microscopes, two-photon confocal microscopes, ultrashort-pulse illumination, and high-aperture objectives. This comprehensive book is based on computer simulations.

E. Hecht, *Optics*, 4th ed., Addison-Wesley, Reading, MA (2001). This is the standard work on optics for the undergraduate level. It offers a clear discussion of geometrical and physical optics. The numerous figures clearly illustrate the fundamental principles of optics.

B. Herman, *Fluorescence Microscopy*, 2nd ed., Springer Verlag, New York (1998). A short review of the fundamentals of fluorescence microscopy.

J. Huisken, J. Swoger, F. Del Bene, J. Wittbrodt, and E. H. K. Stelzer, "Optical sectioning deep inside live embryos by selective plane illumination microscopy," *Science* **305**, 1007–1009 (2004).

J. H. Hunt, Editor, *Selected Papers on Nonlinear Optical Spectroscopy*, SPIE Press, Bellingham, WA (2001). Reprints of full papers on many topics of interest, including two-photon absorption and multiphoton spectroscopy.

S. Inoué and K. R. Spring, *Video Microscopy, The Fundamentals*, 2nd edition, Plenum Press, New York (1997). The first part of this book contains useful chapters on microscope image formation and practical aspects of microscopy. The sections on cameras are clear, but dated in their content.

S. Jutamulia, Editor, *Selected Papers on Near-Field Optics*, SPIE Press, Bellingham, WA (2002).

W. Kaiser and C. G. B. Garrett, "Two-photon excitation in CaF_2:Eu^{2+}," *Physical Review Letters* **7**(6), 229–231 (1961).

F. H. Kasten, "The Origins of Modern Fluorescence Microscopy and Fluorescent Probes," Chapter 1 in *Cell Structure and Function by Microspectrofluorometry*, E. Kohen and J. G. Hirschberg, Editors, 3–50, Academic Press, San Diego, CA (1989). A must-read for those interested in the history and development of fluorescence microscopy.

K. H. Kim, C. Buehler, and P. T. C. So, "High-speed, two-photon scanning microscope," *Applied Optics* **38**, 6004–6009 (1999).

J. R. Lakowicz, *Principles of Fluorescence Spectroscopy*, 2nd ed., Kluwer Academic/Plenum Publishers, New York (1999). This is the standard work for learning about many aspects of fluorescence: time and frequency domain lifetime measurements, fluorescent probes, quenching, anisotropy, energy transfer, and excited state reaction. It describes the theoretical aspects and the instrumentation and analysis of the data.

G. Marriott and I. Parker, *Biophotonics*, Parts A and B, *Methods in Enzymology*, Academic Press (2003). These two volumes review many applications from all types of optical microscopes. It includes a comprehensive chapter on how to construct a confocal microscope.

W. T. Mason, Editor, *Fluorescent and Luminescent Probes for Biological Activity: A Practical Guide to Technology for Quantitative Real-Time Analysis*, 2nd ed., Academic Press, London (1999). A very good source of practical methods using endogenous and exogenous fluorescent probes in microscopy.

B. R. Masters, "The scientific life of Maria Göppert-Mayer," *Optics and Photonics News* **11**(9), 38–41 (2000).

B. R. Masters, Editor, *Selected Papers on Confocal Microscopy*, SPIE Press, Bellingham, WA (1996). A source book of reprinted papers from 1950 to 1996. It also contains key patents from 1884 to 1992 and chronicles the key developments in the development of all types of confocal as well as other types of optically sectioning optical microscopes. Many of the papers and patents cited in this text can be found here.

B. R. Masters and P. T. C. So, Editors, *Handbook of Biological Nonlinear Microscopies*, Oxford University Press (2006).

B. R. Masters and P. T. C. So, "Antecedents of two-photon excitation laser scanning microscopy," *Microscopy Research and Technique* **63**(1), 3–11 (2004). This is a historical overview of the development of nonlinear microscopy.

B. R. Masters, Editor, *Selected Papers on Multiphoton Excitation Microscopy*, SPIE Press, Bellingham, WA (2003). A source book of reprinted papers from 1931 to 2002, it chronicles the key developments in the development of all types of confocal as well as other types of nonlinear optically sectioning optical microscopes. It discusses single-molecule studies and applications to cells, tissues, cell biology, embryology, developmental biology, neuroscience, and dermatology.

B. R. Masters, Editor, *Selected Papers on Optical Low-Coherence Reflectometry and Tomography*, SPIE Press, Bellingham, WA (2001). A source book of reprinted papers that chronicle the key developments of optical low-coherence imaging. There is a balance among theory, instruments, and applications. It includes many applications to cell biology and medicine.

R. C. Mellors and R. Silver, "A microfluorometric scanner for the differential detection of cells: application to exfoliative cytology," *Science* **114**(2962), 356–360 (1951).

D. B. Murphy, *Fundamentals of Light Microscopy and Electronic Imaging*, Wiley-Liss, New York (2001). This is a very good book to learn the fundamentals of microscopy. Each chapter includes practical demonstrations and exercises. It has a good balance between the theory and the practical aspects of optical microscopy. Several laboratory demonstrations of important principles are described.

H. Naora, "Microspectrophotometry and cytochemical analysis of nucleic acids," *Science* **114**, 279–280 (1951). This paper describes the first non-imaging confocal microscope.

T. Nielsen, M. Fricke, D. Hellweg, and P. Andresen, "High efficiency beamsplitter for multifocal multiphoton microscopy," *Journal of Microscopy* **201**(3), 368–376 (2000).

S. W. Paddock, Editor, *Confocal Microscopy Methods and Protocols,* Humana Press, Totowa, NJ (1999). This book contains detailed protocols for confocal imaging of a variety of specimens, from cells to embryos.

J. B. Pawley, Editor, *Handbook of Biological Confocal Microscopy*, Plenum Press, New York (1995). [A new edition will be published in 2006.] This is a very good reference book for both the theory and the applications of biological confocal microscopy.

A. Periasamy, Editor, *Methods in Cellular Imaging*, Oxford University Press, New York (2001). A good source for many biological applications based on confocal microscopy and multiphoton microscopy. This reference is a good source of information on FLIM and FRET microscopic techniques in biological applications.

F. Roberts and J. Z. Young, "The flying-spot microscope," *Proceedings of the IEEE* **99**, Pt. IIIA, 747–757 (1952).

J. Sharpe, U. Ahlgren, P. Perry, B. Hill, A. Ross, J. Hecksher-Sørensen, R. Baldock, and D. Davidson, "Optical projection tomography as a tool for 3D microscopy and gene expression studies," *Science* **296**, 541–545 (2002).

W. T. Silfvast, *Laser Fundamentals*, 2nd ed., Cambridge University Press, Cambridge, UK (2004). A clear description of the physics and design principles of lasers. It describes the major types of lasers and their operation. This is a very complete and up-to-date treatment of lasers.

P. T. C. So, C. Y. Dong, B. R. Masters, and K. M. Berland, "Two-photon excitation fluorescence microscopy," *Annu. Rev. Biomed. Eng.* **2**, 399–429 (2000).

E. M. Slayter and H. S. Slayter, *Light and Electron Microscopy*, Cambridge University Press, New York (1994). This is a good source for an introduction to physical optics and the principles of both light and electron microscopy.

G. M. Svishchev, "Microscope for the study of transparent light-scattering objects in incident light," *Optics and Spectroscopy* **26**, 171–172 (1969).

B. Valeur, *Molecular Fluorescence: Principles and Applications*, Wiley-VCH, Weinheim, Germany, (2002). A highly recommended book on the physical basis of fluorescence and practical applications of fluorescence.

T. Wilson, Editor, *Confocal Microscopy*, Academic Press, London (1990). This book is edited by one of the developers of confocal microscopy. It is a good source for theory and applications of confocal microscopy.

T. Wilson and C. Sheppard, *Theory and Practice of Scanning Optical Microscopy*, Academic Press, London (1984). This is the classic book on confocal microscopy. It develops and uses Fourier image theory that is applied to the optical properties of various instruments. Applications to semiconductor and integrated circuits as well as nonlinear optical microscopy, super-resolution, and direct-view scanning microscopy are covered.

R. Yuste and A. Konnerth, *Imaging in Neuroscience and Development, A Laboratory Manual*, Cold Spring Harbor Laboratory Press, Cold Spring Harbor, New

York (2005). This is a practical guide for the imaging of tissues and organisms of key importance for neuroscience and development. The tutorial on microscopy and microscope optical systems by Lanni and Keller is both clear and comprehensive.

R. Yuste, F. Lanni, and A. Konnerth, Editors, *Imaging Neurons, A Laboratory Manual,* Cold Spring Harbor Laboratory Press, Cold Spring Harbor, NY (2000). A good book for fundamentals of optical microscopy, confocal microscopy, and multiphoton microscopy; there are many applications to *in vivo* microscopy in the field of neurobiology; however, the material is useful for other *in vivo* applications.

Journals

Applied Optics
Biophysical Journal
Journal of Biomedical Optics
Journal of Microscopy
Journal of Optical Society of America
Microscopy Research and Technique
Optics Communications
Optics Express
Optics Letters

Special Journal Issues on Multiphoton Microscopy

A. Periasamy and A. Diaspro, Editors, *Journal of Biomedical Optics* **8**(3), July 2003.

A. Diaspro, Editor, *Microscopy Research and Technique* **63**(1), 2004.

A. Diaspro, Editor, *Microscopy Research and Technique* **47**(2), 1999.

Internet Resources

Fluorescent Probes

Molecular Probes, Inc.: http://probes.invitrogen.com/

This website contains links to many other Web resources: noncommercial, journal, commercial sites, conferences, and other meetings.

Their online catalog contains a useful tutorial on many aspects of fluorescence and an extensive catalog of fluorescence probes for labeling: ions, molecules, cells, tissues and organs. Their catalog is actually a wonderful, comprehensive reference containing application images, references, absorption and emission spectral data, and the chemical and photochemical properties of all their products under a variety of conditions. Detailed protocols are provided for loading cells, calibrating fluorescence intensity, the use of caging groups and their photolysis, the study of signal transduction, using potentiometric probes, and using dyes to determine ion concentration and pH. They also have books on microscopy and fluorescence techniques and a good variety of calibration systems. The protocols contained in the handbook cover the scale from membranes, cell organelles, cells, tissues, and organs to whole organisms used for studies of their developmental biology. The online version of the handbook is updated often and of great utility.

Information on confocal microscopy, multiphoton excitation microscopy, microscopes, lasers, image processing software, techniques, technical information on microscope objectives, light sources, microscope images from many types of microscopes, and a wide range of detailed technical application notes can be found here.

Quantum Dot Fluorescent Probes, Quantum Dot Corporation, Hayward, CA: http://qdots.com

Microscopes and Tutorials on Microscopy

Florida State University's Molecular Expressions: http://micro.magnet.fsu.edu
Includes confocal and multiphoton microscopy Java tutorials.

Leica: http://www.leica-microsystems.com

Nikon Instruments, Inc.: http://www.nikonusa.com/, http://www.microscopyu.com
Interactive Java tutorials.

Olympus: www.olympusmicro.com, www.olympusamerica.com
Interactive Java tutorials and Microscopy Resource Center, which contains a section on microscopy history, several websites showing collections of antique microscopes from around the world.

Carl Zeiss: www.zeiss.com, www.zeiss.de/lsm
Tutorials and application notes on all aspects of microscopy; searchable database.

Lasers

Coherent, Inc.: http://www.cohr.com

Other Components

Physikinstrumente: www.physikinstrumente.de
Tutorial on piezoelectrics in micropositioning devices; piezoelectricity and piezo actuators.

Newport Corporation and Spectra Physics: http://www.newport.com
A good source of CW and pulsed lasers and an optics tutorial.

New Focus: http://www.newfocus.com

Detectors

Hamamatsu: http://www.hamamatsu.com

Becker & Hickl GmbH, Berlin, Germany: http://www.becker-hickl.de/

Photometrics, a division of Roper Scientific, Inc.: www.roperscientific.com
This website contains technical information and application notes for cooled, back-illuminated, high quantum efficiency (90%) charge-coupled-device cameras with on-chip multiplication gain.

Scanners

Cambridge Technology, Inc.: www.camtech.com

Optical Filters

Omega Optical Inc.: http://www.omegafilters.com

Chroma Technology Corporation: http://www.chroma.com

Microscopy Societies

Microscopy Society of America: http://www.microscopy.org

Royal Microscopical Society: http://www.rms.org.uk/

Image Processing Software

Free image processing software "ImageJ" for many computer platforms, http://rsb.info.nih.gov/ij/

Patents

United States Patent and Trademark Office: www.uspto.gov.
Patents are an excellent source of information for the understanding, design, and construction of instruments. Here you can search by key words, patent inventor name, patent number.

Other Websites

Professor Peter So laboratory: http://web.mit.edu/solab/
This website provides a wealth of information on the engineering of novel microscopy instrumentation and the application of these new tools to biomedical problems. These new types of microscopic and spectroscopic instruments are designed to span the range from single molecule dynamics, to the cellular level, to the tissue level. There are useful links for optical instrumentation.

Professor Stefan Hell laboratory: www.4pi.de
Tutorial information on how to overcome the Abbe diffraction limit in light microscopy and achieve three-dimensional resolution in the 100 nm range. The group's publications are available as PDFs. There are links to sites on the history of the microscope.

References for Applications in Ophthalmology and Dermatology

Ophthalmology

These references cover the development of instruments and the *ex vivo* and *in vivo* microscopic investigation of cells, tissues, and organs. There are instruments designed to use light microscopy to monitor cellular metabolism; optical techniques to provide three-dimensional microscopy of the cornea, the ocular lens, and the optic nerve *in vivo*; the development of clinical confocal microscopes for diagnostic "optical biopsy" of the living eye; the use of confocal microscopy to investigate redox metabolism is developed, as is the use of multiphoton excitation microscopy to monitor redox metabolism in the *ex vivo* cornea; correlative microscopy is demonstrated by the use of both confocal and electron microscopy on the same human lenses in the same regions.

B. R. Masters, "Noninvasive redox fluorometry: how light can be used to monitor alterations of corneal mitochondrial function," *Current Eye Research* **3**, 23–26 (1984).

B. R. Masters, "Effects of contact lenses on the oxygen concentration and epithelial redox state of rabbit cornea measured noninvasively with an optically sectioning redox fluorometer microscope,"in *The Cornea: Transactions of the World Congress on the Cornea III*, H.D. Cavanagh, Editor, Raven Press, New York, 281–286 (1988).

B. R. Masters and S. Paddock, "In vitro confocal imaging of the rabbit cornea," *J. Microscopy* **158**(2), 267–274 (1990).

G. Q. Xiao, G. S. Kino, and B. R. Masters, "Observation of the rabbit cornea and lens with a new real-time confocal scanning optical microscope," *Scanning* **12**(3), 161–166 (1990).

B. R. Masters and S. W. Paddock, "Three-dimensional reconstruction of the rabbit cornea by confocal scanning optical microscopy and volume rendering," *Applied Optics* **29**, 3816–3822 (1990).

B. R. Masters, "Two and three-dimensional visualization of the living cornea and ocular lens," *Machine Vision and Applications*, Special Issue on Three-Dimensional Microscopy **4**, 227–232 (1991).

B. R. Masters, "Confocal microscopy of the *in situ* crystalline lens," *Journal of Microscopy* **165**, 159–167 (1992).

B. R. Masters, A. Kriete, and J. Kukulies, "Ultraviolet confocal fluorescence microscopy of the *in vitro* cornea: redox metabolic imaging," *Applied Optics* **34**(4), 592–596 (1993).

B. R. Masters, "Specimen preparation and chamber for confocal microscopy of the eye," *Scanning Microscopy* **7**(2), 645–651 (1993).

B. R. Masters, M. A. Farmer, "Three-dimensional confocal microscopy and visualization of the *in situ* cornea," *Computerized Medical Imaging and Graphics* **17**(3), 211–219 (1993).

F. W. Fitzke and B. R. Masters, "Three-dimensional visualization of confocal sections of *in vivo* human fundus and optic nerve," *Curr. Eye Res.* **12**, 1015–1018 (1993).

B. R. Masters and A. A. Thaer, "Real-time scanning slit confocal microscopy of the *in vivo* human cornea," *Applied Optics* **33**(4), 695–701 (1994).

B. R. Masters and A. A. Thaer, "*In vivo* human corneal confocal microscopy of identical fields of subepithelial nerve plexus, basal epithelial and wing cells at different times," *Microscopy Research and Techniques* **29**, 350–356 (1994).

B. R. Masters, "Scanning slit confocal microscopy of the *in vivo* cornea," *Optical Engineering* **34**(3), Feature Issue on Optical Engineering in Ophthalmology, S. Jutamulia and T. Asakura, Editors,684–692 (1995).

D. W. Piston, B. R. Masters, and W. W. Webb, "Three-dimensionally resolved nad(p)h cellular metabolic redox imaging of the *in situ* cornea with two-photon excitation laser scanning microscopy," *J. Microscopy* **178**, 20–27 (1995).

B. R. Masters and A. A. Thaer, "*In vivo*, real-time confocal microscopy of the continuous wing cell layer adjacent to the basal epithelium in the human cornea: a new benchmark for *in vivo* corneal microscopy," *Bioimages* **3**(1), 7–11 (1995).

D. C. Beebe and B. R. Masters, "Cell lineage and the differentiation of corneal epithelial cells," *Invest Ophthalmol Vis Sci.* **37**(9), 1815–1825 (1996).

B. R. Masters, K. Sasaki, Y. Sakamoto, M. Kojima, Y. Emori, S. Senft, and M. Foster, "Three-dimensional volume visualization of the *in vivo* human ocular lens showing localization of the cataract," *Ophthalmic Research.* **28**, suppl. 1, 120–126 (1996).

B. R. Masters, "Three-dimensional confocal microscopy of the lens," *Ophthalmic Research* **28**, suppl. 1, 115–119 (1996).

M. Böhnke and B. R. Masters, "Long-term contact lens wear induces a corneal degeneration with micro-dot deposits in the corneal stroma," *Ophthalmology* **104**, 1887–1896 (1997).

B. R. Masters, G. F. J. M. Vrensen, B. Willekens, and J. Van Marle, "Confocal light microscopy and scanning electron microscopy of the human eye lens," *Exp. Eye Res.* **64**(3), 371–377 (1997).

F. W. Fitzke, B. R. Masters, R. J. Buckley, and L. Speedwell, "Fourier transform analysis of human corneal endothelial specular photomicrographs," *Exp. Eye Res.* **65**, 205–214 (1997).

B. R. Masters and A. A. Thaer, "Real-time confocal microscopy of *in vivo* human corneal nerves," *Bioimages* **4**(3), 129–134 (1997).

B. R. Masters and S. L. Senft, "Transformation of a set of slices rotated on a common axis to a set of z-slices: application to three-dimensional visualization of the *in vivo* human lens," *Computerized Medical Imaging and Graphics* **21**(3), 145–151 (1997).

B. R. Masters, "Three-dimensional microscopic tomographic imaging of the cataract in a human lens *in vivo*," *Optics Express* **3**(9), 332–338 (1998). URL: http://www.opticsexpress.org

B. R. Masters, "Three-dimensional confocal microscopy of the living *in situ* rabbit cornea," *Optics Express* **3**(9), 351–355 (1998). URL: http://www.opticsexpress. org

B. R. Masters, "Three-dimensional confocal microscopy of the human optic nerve *in vivo*," *Optics Express* **3**(10), 356–359 (1998). URL: http://www.opticsexpress.org

B. R. Masters and B. Chance, "Redox confocal imaging: intrinsic fluorescent probes of cellular metabolism,"in *Fluorescent and Luminescent Probes,*2nd Edition, W. T. Mason, Editor, Academic Press, London, UK, 361–374 (1999).

B. R. Masters and M. Böhnke, "Video-rate, scanning slit, confocal microscopy of the living human cornea *in vivo*: three-dimensional confocal microscopy of the eye," *Methods in Enzymology*, **307**, *Confocal Microscopy*, P. Michael Conn, Editor, Academic Press, New York, 536–563 (1999).

M. Böhnke and B. R. Masters, "Confocal microscopy of the cornea," *Progress in Retina & Eye Research* **18**(5), 553–628 (1999).

B. R. Masters and M. Böhnke, "Three-dimensional confocal microscopy of the living eye," *Annual Review of Biomedical Engineering* **4**, Annual Reviews, Palo Alto, CA, 69–91 (2002).

B. R. Masters, "David Maurice's contribution to optical ophthalmic instrumentation: roots of the scanning-slit clinical confocal microscope," *Experimental Eye Research* **78**, 315–326 (2004).

Dermatology

These papers demonstrate the use of in vivo confocal microscopy and in vivo multiphoton excitation microscopy and spectroscopy to investigate the structure and function of in vivo human skin.

B. R. Masters, "Three-dimensional confocal microscopy of human skin *in vivo*: autofluorescence of normal skin," *Bioimages* **4**(1), 1–7 (1996).

B. R. Masters, D. Aziz, A. Gmitro, J. Kerr, B. O'Grady, and L. Goldman, "Rapid observation of unfixed, unstained, human skin biopsy specimens with confocal microscopy and visualization," *Journal of Biomedical Optics* **2**(4), 437–445 (1997).

B. R. Masters, G. Gonnord, and P. Corcuff, "Three-dimensional microscopic biopsy of *in vivo* human skin: a new technique based on a flexible confocal microscope," *Journal of Microscopy* **185**(3), 329–338 (1997).

B. R. Masters, P. T. C. So, and E. Gratton, "Multiphoton Excitation Fluorescence Microscopy and Spectroscopy of *in vivo* Human Skin," *Biophysical Journal* **72**, 2405–2412 (1997).

B. R. Masters, P. T. C. So, and E. Gratton, "Multiphoton excitation microscopy and spectroscopy of cells, tissues, and human skin *in vivo*," *Fluorescent and Luminescent Probes,* 2nd Edition, W. T. Mason, Editor, Academic Press, London, UK, 414–432 (1999).

B. R. Masters, P. T. C. So, K. Kim, C. Buehler, and E. Gratton, "Multiphoton Excitation Microscopy, Confocal Microscopy, and Spectroscopy of Living Cells and Tissues; Functional Metabolic Imaging of Human Skin *in vivo*," *Methods in Enzymology* **307**, *Confocal Microscopy*, P. Michael Conn, Editor, Academic Press, New York, 513–536 (1999).

B. R. Masters and P. T. C. So, "Multiphoton excitation microscopy of human skin *in vivo*: early development of an optical biopsy," *Saratov Fall Meeting '99: Optical Technologies in Biophysics and Medicine*, V. V. Tuchin, D. A. Zimnyakov, and A. B. Pravdin, Editors, *SPIE Proc.* **4001**, 156–164 (2000).

B. R. Masters and P. T. C. So, "Confocal microscopy and multiphoton excitation microscopy of human skin *in vivo*," *Optics Express* **8**(1), 2–10 (2001).

B. R. Masters, P. T. C. So, C. Buehler, N. Barry, J. D. Sutin, W. M. Mantulin, and E. Gratton, "Mitigating thermal-damage potential during two-photon dermal imaging," *Journal of Biomedical Optics* **9**(6), 1265–1270 (2004).

Index

Barry R. Masters, formerly a Gast Professor in the Department of Ophthalmology, University of Bern, Switzerland, is currently an independent consultant. He was a professor at the Uniformed Services University of the Health Sciences in Bethesda, Maryland. He is a Fellow of both the Optical Society of America (OSA) and SPIE—The International Society for Optical Engineering. He received a BSc degree in Chemistry from the Polytechnic Institute of Brooklyn, an MSc degree in Physical Chemistry from Florida State University (Institute of Molecular Biophysics), and a Ph.D. degree in Physical Chemistry from the Weizmann Institute of Science in Israel. He is an editor or author of several books: *Noninvasive Diagnostic Techniques in Ophthalmology* (1990); *Medical Optical Tomography: Functional Imaging and Monitoring* (1993); *Selected Papers on Confocal Microscopy* (1996); *Selected Papers on Optical Low-Coherence Reflectometry and Tomography* (2000); and *Selected Papers on Multiphoton Excitation Microscopy* (2003). He is a co-editor of *Biomedical Optical Biopsy*, an OSA CD-ROM. He has published 80 refereed research papers, 110 book chapters and proceedings, and 105 scientific abstracts. In 1999 Professor Masters and Professor Böhnke shared the Vogt Prize for Research (the highest Swiss award for Ophthalmology) for their research on the confocal microscopy of the cornea. He received an AAAS Congressional Science & Engineering Fellowship (OSA/SPIE) in 1999–2000. Dr. Masters has been a Visiting Professor at The Netherlands Ophthalmic Research Institute, Amsterdam; Beijing Medical University, Beijing, PRC; Science University of Tokyo, Japan; University of Bern, Switzerland; and a Visiting Research Fellow, Nuffield Laboratory of Ophthalmology, University of Oxford. He is a member of the editorial board of *Computerized Medical Imaging and Graphics; Graefe's Archive for Clinical and Experimental Ophthalmology;* and *Ophthalmic Research.* He is a member of OSA's Applied Optics Patent Review Panel. His research interests include the development of *in vivo* confocal microscopy of the human eye and skin, cell biology of differentiation and proliferation in epithelial tissues, the application and development of multiphoton excitation microscopy to deep-tissue imaging and spectroscopy, one- and two-photon metabolic redox imaging, diagnostic and functional medical imaging, optical Fourier transform methods for cellular pattern recognition, and fractal analysis of the vascular system.